# SUPER
# SNEAKY
# USES FOR
# EVERYDAY
# THINGS

## Other Books in the Sneaky Science Series

# SUPER
# SNEAKY
# USES FOR
# EVERYDAY
# THINGS

Power Devices with Your Plants,
Modify High-Tech Toys, Turn a Penny into
a Battery, Make Sneaky Light-Up Nails and
Fashion Accessories, and Perform Sneaky
Levitation with Everyday Things

## CY TYMONY

Andrews McMeel
Publishing, LLC

Kansas City • Sydney • London

Andrews McMeel Publishing, LLC
an Andrews McMeel Universal company
1130 Walnut Street, Kansas City, Missouri 64106

www.andrewsmcmeel.com

11 12 13 14 15 RR2 10 9 8 7 6 5 4 3 2 1

ISBN: 978-1-4494-0814-5

Library of Congress Control Number: 2011926180

**Attention: Schools and Businesses**

Andrews McMeel books are available at quantity discounts with bulk purchase for educational, business, or sales promotional use. For information, please e-mail the Andrews McMeel Publishing Special Sales Department:
specialsales@amuniversal.com

# DISCLAIMER

This book is for the entertainment and edification of its readers. While reasonable care has been exercised with respect to its accuracy, the publisher and the author assume no responsibility for errors or omissions in its content. Nor do we assume liability for any damages resulting from use of the information presented here.

This book contains references to electrical safety that *must* be observed. *Do not use AC power for any projects listed.* Do not place or store magnets near such magnetically sensitive media as videotapes, audiotapes, or computer disks.

Disparities in materials and design methods and the application of the components may cause results to vary from those shown here. The publisher and the author disclaim any liability for injury that may result from the use, proper or improper, of the information contained in this book. We do not guarantee that the information contained herein is complete, safe, or accurate, nor should it be considered a substitute for your good judgment and common sense.

Nothing in this book should be construed or interpreted to infringe on the rights of other persons or to violate criminal statutes. We urge you to obey all laws and respect all rights, including property rights, of others.

# CONTENTS

# PART 2

## Sneaky Fashions    27

# PART 3

## Sneaky Toys and Games    53

# PART 4

## Sneaky Science Projects   93

## Bonus Sneaky Projects   119

# ACKNOWLEDGMENTS

I'd like to thank my agents, Sheree Bykofsky and Janet Rosen, for believing from the start in my *Sneaky Uses* book concept. Special thanks to Katie Anderson and Dorothy O'Brien, my Andrews McMeel editors, for their valuable support in shaping the book series. And a thank you to the book's illustrator, Kevin Brimmer.

I'm grateful for the project evaluation and development assistance provided by Sybil Smith and Bill Melzer.

And a special thanks to Helen Cooper, Clyde Tymony, George and Zola Wright, David Townsend and family, Ronald Mitchell, Kevin Burnley, and to my mother, Cloise, for providing positive motivation and support for an early foundation in science, and a love of reading.

# INTRODUCTION

People rarely think about the common items they use in everyday life. Sure, they serve their purpose, but can they be used in any other ways? For many, the answer is a resounding, "Yes!" You can perform amazing feats with ordinary items around you and with no special knowledge. How? By thinking differently about those items, to make *sneaky* uses of everyday things.

A *sneaky* person is curious enough to take things apart, examine them, and combine them with other items for fun and profit—and sometimes get themselves and others out of a jam. When you learn sneaky reuse techniques you also save money, become more self-reliant, and can pass on amazing techniques to your friends and family.

Did you know that you can turn a grocery bag into an air cannon? Or generate electricity with plants or a pencil? In a pinch, do you know what else a 9-volt battery can substitute for?

*Super Sneaky Uses for Everyday Things* provides you with these techniques and a lot more. In it you'll find innovative, do-it-yourself fashion crafts, sneaky games, and science projects using items you already have around the house. You'll never look at "waste" in the same way again. With just a few easily obtainable supplies, you will have the makings of a personal sneaky craft lab.

Each of the projects in this book is designed to teach you how everyday items work and how they can be reused in practical and fun applications. The chapters that follow are crammed with more than thirty crafts and projects. From simple paper clip applications to unique techno-fashion innovations and toy modifications, these sneaky constructs are just the ticket for innovative fun.

One suggestion: Review this entire book before you start any project. The book is organized to help you understand and build on your sneaky skills from simple tricks all the way up to more complex gadget adaptations. And don't be intimidated; it's easy. You'll have a lot of fun thinking about the various crafts you can make and how to go even further with your own creations. You'll probably even modify some of the projects you learn and make them even sneakier!

Studying the "Sneaky Fundamentals" section will give you an idea of what amazing new applications you can find for everyday objects. You'll learn sneaky sources for wire and how to connect things. And there's a special "Sneaky Uses for a Paper Clip" section that you should not miss.

Professional crafters will enjoy *Super Sneaky Uses for Everyday Things* too. You'll find innovative techniques that will allow you to make items not seen anywhere else. All the projects are tested safe and can be completed in a short time.

After you make a project or two from this book, you'll be inspired to teach others your sneaky knowledge. Turning your discards into fun toys and making unique gifts for others is another benefit to discovering *Super Sneaky Uses for Everyday Things.*

Let's get started!

# SNEAKY FUNDAMENTALS

# SNEAKY WIRE SOURCES

Ordinary wire can be used in many sneaky ways. You'll soon learn how it can be utilized to make a radio transmitter, a speaker, and more. When wire is required for sneaky projects, whenever possible try to use everyday items that you might otherwise have thrown away. Recycling metal will help save our natural resources.

## Getting Wired

In an emergency, you can obtain wire—or items that can be used as wire—from some very unlikely sources. **Figure 1** illustrates just a few of the possible items that you can use in case connecting wire is not available.

Ready-to-use wire can be obtained from:

- ▶ Telephone cords
- ▶ TV/VCR cables
- ▶ Headphone wire
- ▶ Earphone wire
- ▶ Speaker wire
- ▶ Wire from inside toys, radios, and other electrical devices

**Note:** Some of the sources above will have one to six separate wires inside.

Wire for projects can also be made from:

- ▶ Take-out food container handles
- ▶ Twist-ties
- ▶ Paper clips
- ▶ Envelope clasps
- ▶ Ballpoint pen springs
- ▶ Fast-food wrappers
- ▶ Potato chip bag liners

You can also use aluminum from the following items:

- ▶ Margarine wrappers
- ▶ Ketchup and condiment packages
- ▶ Breath mint container labels
- ▶ Chewing gum wrappers
- ▶ Trading card packaging
- ▶ Coffee creamer container lids

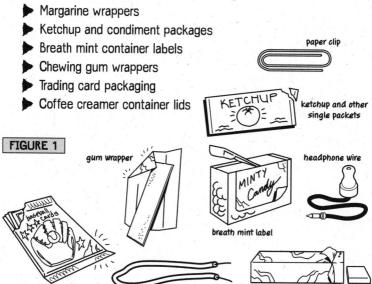

FIGURE 1

paper clip

ketchup and other single packets

gum wrapper

headphone wire

breath mint label

trading card package

TV/VCR cables

margarine wrapper

**Note:** The wires used from the sources above are only to be used for low-voltage, battery-powered projects.

Use special care when handling fragile aluminum materials. In some instances, aluminum may be coated with a wax or plastic coating that you may be able to remove.

You can cut strips of aluminum material from food wrappers easily enough. With smaller items—such as aluminum obtained from a coffee cream container—use the sneaky cutting pattern shown in **Figure 2**.

Making resourceful use of items to make sneaky wire is not only intriguing, it's fun.

**FIGURE 2**

SNEAKY COFFEE CREAMER WIRE

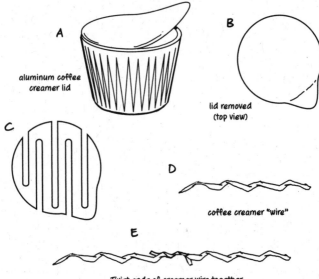

A

aluminum coffee
creamer lid

B

lid removed
(top view)

C

D

coffee creamer "wire"

E

Twist ends of creamer wire together
for longer runs.

# HOW TO CONNECT THINGS

The "Getting Wired" project illustrated how to obtain wire from everyday things. Now you'll learn how to connect the wires to provide consistent performance. (A tight connection is crucial to the operation of electrical projects, otherwise faulty and erratic results may occur.)

**Figure 1** shows a piece of insulated wire. The insulation material must be stripped away to make a metal-to-metal connection to other electrical parts. Strip away about one to two inches of insulation from both ends of the wire. See **Figure 2**.

To connect the wire to another wire lead, wrap both ends around each other, as shown in **Figure 3**.

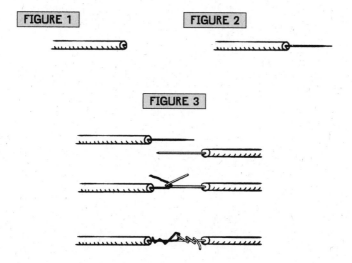

When connecting the wire to the end of a stiff lead (such as the end of an LED), wrap the wire around the lead and bend the lead back over the wire. See **Figure 4**.

To connect the wire to the end of a small battery, bend the wire into a circular shape, place it on the battery terminal, and wrap the connection tightly with tape, as shown in **Figure 5**.

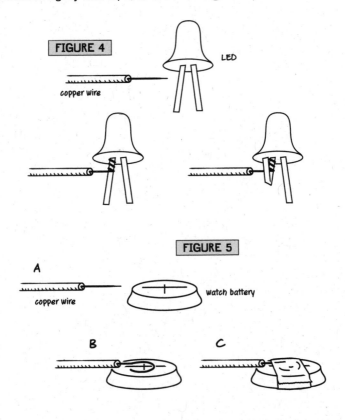

FIGURE 4

LED

copper wire

FIGURE 5

A

copper wire

watch battery

B

C

# ELECTRICITY FUN-DAMENTALS

Many forms of alternative energy, including wind, thermal, hydro, even nuclear, are used to generate electrical power. This is accomplished by moving blades (wind or hydro) or heating water into steam (thermal or nuclear) to turn an electrical generator. The following projects illustrate how electrical power is produced and the relationship between electricity and magnetism.

## WHAT'S NEEDED

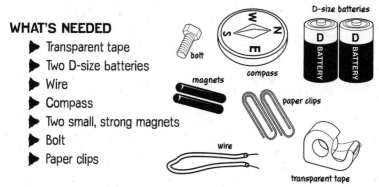

- ▶ Transparent tape
- ▶ Two D-size batteries
- ▶ Wire
- ▶ Compass
- ▶ Two small, strong magnets
- ▶ Bolt
- ▶ Paper clips

## WHAT TO DO

When electricity flows through a wire, a magnetic field is produced around it. To test this, first tape the two D-size batteries together and place them near a length of wire. Next, set the compass near the wire and hold the ends of the wire to both battery terminals (only for a few seconds), and you'll see the compass pointer move. See **Figure 1**.

Position the wire vertically in a small loop shape and touch the battery terminals. Bring a magnet close to the wire as you connect and disconnect the wire, and you'll see the wire move because it has become an electromagnet. See **Figure 2**.

If you wrap the wire thirty times around a bolt and connect it to the battery terminals, it will become an electromagnet. **Figure 3** shows how it can attract and lift paper clips.

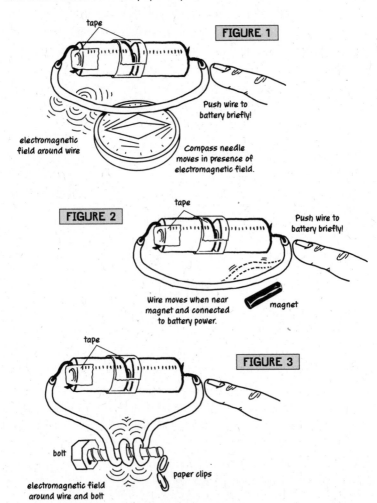

tape

**FIGURE 1**

Push wire to
battery briefly!

electromagnetic
field around wire

Compass needle
moves in presence of
electromagnetic field.

**FIGURE 2**

tape

Push wire to
battery briefly!

Wire moves when near
magnet and connected
to battery power.

magnet

tape

**FIGURE 3**

bolt

paper clips

electromagnetic field
around wire and bolt

# SIMPLE ELECTRICAL CIRCUITS

To test an item's conductivity (the ability to let electricity flow through it), use a flashlight bulb or an LED (light-emitting diode). An LED is used in most electronic devices and toys as a function indicator because it draws very little electrical current, operates with very little heat, and has no filament that would burn out.

Lay a 3-volt watch battery on the item, as shown in **Figure 1**. If the lightbulb or LED lights, then the item can be used as wire for battery-powered projects.

**Note:** If the bulb or LED does not light, reverse the connections of its leads and test it again. LEDs are polarity (direction) sensitive.

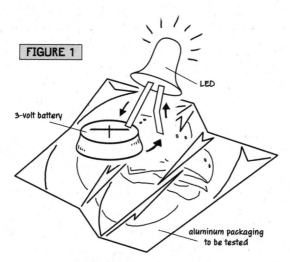

**FIGURE 1**

LED

3-volt battery

aluminum packaging
to be tested

Current flows from the battery's negative (-) terminal to the aluminum
foil to the LED to the battery's positive (+) terminal in a circle.

**Figure 2** shows a simple electrical circuit that consists of a battery, connecting wire, and a lightbulb. Power flows in a circle (*circuit* means "circle") from the negative battery terminal to the light and back to the positive battery terminal.

When using one 1½-volt battery in a circuit, you must use a lightbulb rated as the same voltage. This applies to whatever else you may want to turn on, such as a buzzer or motor.

LEDs generally require two to three volts (unless otherwise noted) to turn on, so connect two 1½-volt batteries in series (end to end) to activate an LED.

**Figure 3** illustrates how to do this. If the LED does not turn on, reverse its leads and test it again.

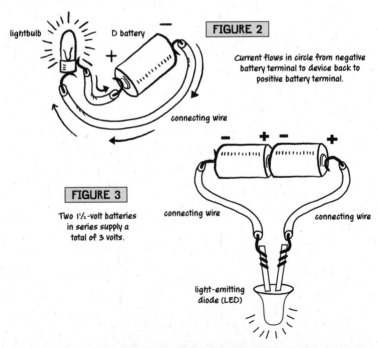

lightbulb   D battery

**FIGURE 2**

Current flows in circle from negative battery terminal to device back to positive battery terminal.

connecting wire

**FIGURE 3**

Two 1½-volt batteries in series supply a total of 3 volts.

connecting wire

connecting wire

light-emitting diode (LED)

# HOW TO BE RESOURCEFUL: SNEAKY USES FOR A PAPER CLIP

You can perform amazing feats with ordinary items without special knowledge or skills. The key is to think outside the box—to see things as what they can become and not just what you think they're limited to be.

For example, a paper clip is designed to hold papers together, but is that all it's good for? Can you reshape it for other practical purposes? Yes, you can!

Consider the following three-step illustrated examples using jumbo paper clips:

## SECRET MESSAGE ETCHER
1. Straighten a paper clip.
2. Lightly etch a secret message or code on paper with the ends of the paper clip. Conceal the paper between two other pieces of paper.
3. Your friend can read the secret message by rubbing the paper with a pencil.

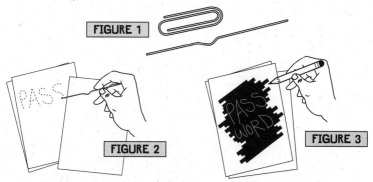

FIGURE 1

FIGURE 2

FIGURE 3

## STATIC ELECTRICITY DETECTOR

1. Bend a paper clip into the shape shown in **Figure 1**.
2. Hang it on the side of a cup. Place little thin strips of aluminum foil on the end of the paper clip hook inside the cup.
3. Rub a balloon or piece of Styrofoam on your hair or sweater and touch it to the end of the paper clip and you'll see the aluminum strips flutter, indicating a high static charge.

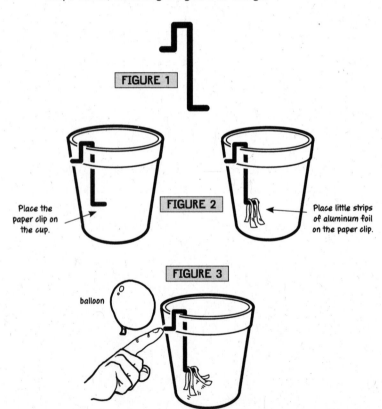

FIGURE 1

Place the paper clip on the cup.

FIGURE 2

Place little strips of aluminum foil on the paper clip.

FIGURE 3

balloon

## COTTER PIN

1. Bend a paper clip into a flattened **C** shape.
2. Then bend it into the shape shown in **Figure 2**.
3. Stick the paper clip into the hole in the nut and bolt and bend the ends back to ensure that the nut will not slip off.

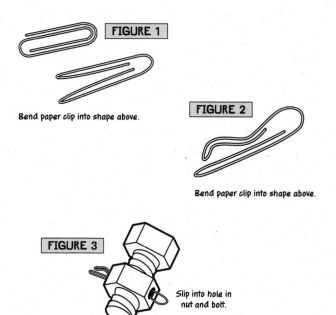

**FIGURE 1**

Bend paper clip into shape above.

**FIGURE 2**

Bend paper clip into shape above.

**FIGURE 3**

Slip into hole in nut and bolt.

## SAILBOAT

1.  Bend one end of the paper clip upward and straight.
2.  Cut a piece of paper into a triangle shape and poke two small holes in it.
3.  Slip the paper on the paper clip, gently set it on the water's surface, and watch it float.

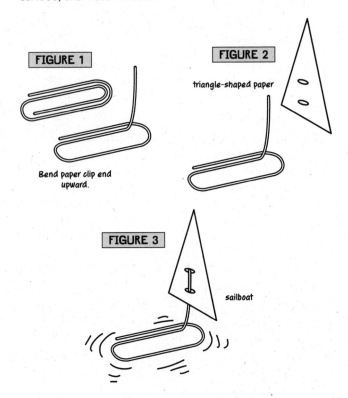

**FIGURE 1**

Bend paper clip end upward.

**FIGURE 2**

triangle-shaped paper

**FIGURE 3**

sailboat

## DISC DRIVE OPENER

1. Straighten a paper clip.
2. Locate the small door release hole on the disc drive.
3. Push the end of a paper clip into the release hole to open the door.

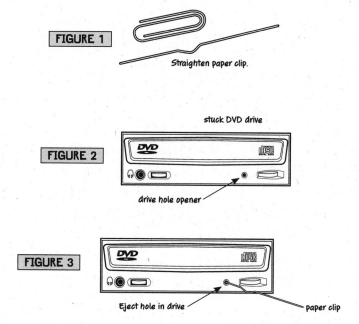

FIGURE 1

Straighten paper clip.

stuck DVD drive

FIGURE 2

drive hole opener

FIGURE 3

Eject hole in drive

paper clip

## MAGNIFIER

1. Bend one end of a paper clip into a loop.
2. Dip the loop into water.
3. Look through the water droplet to view an enlarged image.

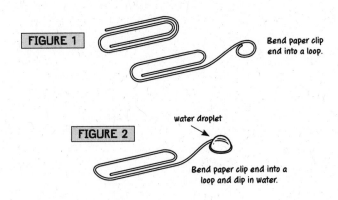

FIGURE 1

Bend paper clip
end into a loop.

water droplet

FIGURE 2

Bend paper clip end into a
loop and dip in water.

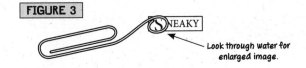

FIGURE 3

SNEAKY

Look through water for
enlarged image.

## COMPASS

1. Straighten a paper clip.
2. Rub it a dozen times in one direction across a magnet, such as the one on the back of a speaker. (If you don't have a magnet, rub the paper clip 200 times in one direction across a piece of silk or acrylic material to magnetize it.)
3. Poke the paper clip into a small piece of coated paper or a leaf. It will float in water, pointing north and south.

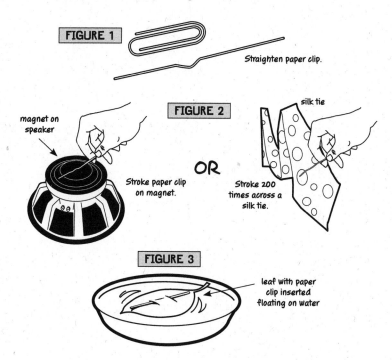

**FIGURE 1**
Straighten paper clip.

**FIGURE 2**
magnet on speaker
Stroke paper clip on magnet.

OR

silk tie
Stroke 200 times across a silk tie.

**FIGURE 3**
leaf with paper clip inserted floating on water

## CONNECTING WIRE

1. Bend the paper clip as shown in **Figure 1**.
   Make a hooked loop on one end.
2. Slide a 1½-volt bulb into the looped end of the paper clip.
3. Hold the battery firmly between the bulb and other end of the paper clip to make a sneaky flashlight.

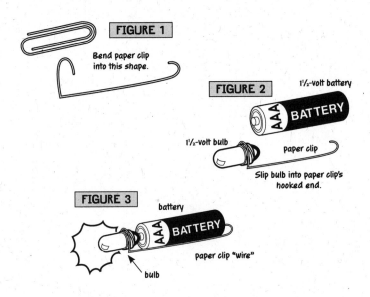

**FIGURE 1**

Bend paper clip
into this shape.

**FIGURE 2**

1½-volt battery

AAA BATTERY

1½-volt bulb

paper clip

Slip bulb into paper clip's
hooked end.

**FIGURE 3**

battery

AAA BATTERY

paper clip "wire"

bulb

## BATTERY ELECTRODE

1. Straighten a paper clip.
2. Stick a piece of copper wire into a fruit such as a lemon.
3. Stick the paper clip into the fruit, and voltage will be produced, as displayed on a voltmeter.

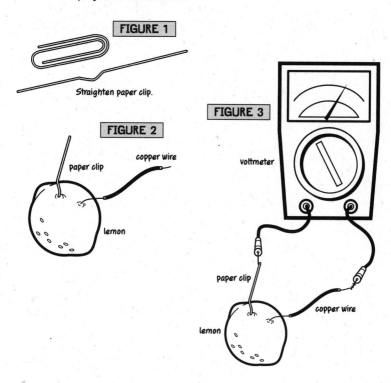

FIGURE 1

Straighten paper clip.

FIGURE 2

paper clip

copper wire

lemon

FIGURE 3

voltmeter

paper clip

copper wire

lemon

## AUDIOCASSETTE REWINDER

1. Straighten a paper clip.
2. Bend the paper clip into the shape shown in **Figure 2,** with a **C** bend near one end.
3. Press the curved part of the paper clip into the audiocassette reel (reshape the paper clip until it's snug) and crank the tape reel around to rewind it.

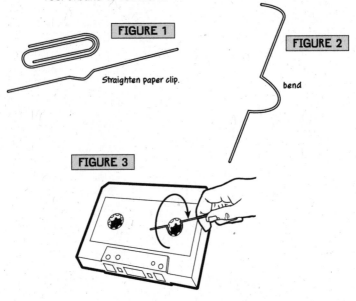

**FIGURE 1**

Straighten paper clip.

**FIGURE 2**

bend

**FIGURE 3**

## MAGICAL LEVITATOR

1. Bend a large paper clip into a stand shape.
2. Tie a small length of (nearly) invisible nylon thread around the bottom.
3. Tie the other end of the thread to a small paper clip. Place a small magnet at the top corner of the jumbo paper clip. Hold the small paper clip up near the magnet (if it's too close, wrap it around the jumbo paper clip until it is just out of reach). The paper clip will try to reach the magnet, but the thread holds it back, and it levitates like magic. You can cover the small paper clip with a small printed picture to make it levitate in midair.

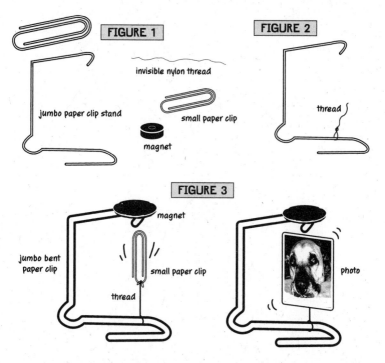

## SNEAKY RING

1. Bend the paper clip into the shape shown in **Figure 1**. Notice the small hooks at the end.

2. Bend the clip until it's a perfect circle and slip the hook ends into the other end.

3. Slip on the ring. Optional: Place a magnet on top of the ring, fold a bill in half, and bring it close to the ring. The currency will move toward the magnetic ring because of the metal particles in the ink.

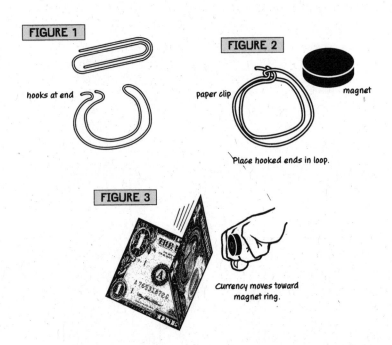

FIGURE 1

hooks at end

FIGURE 2

paper clip

magnet

Place hooked ends in loop.

FIGURE 3

Currency moves toward magnet ring.

## MAGNETIC SWITCH

1. Bend a paper clip into the shape shown in **Figure 1**. Notice the hook on one end.

2. Tape a 1½-volt bulb and an AA battery to a piece of cardboard as shown in **Figure 2**. Place the paper clip's hook on the positive (+) protrusion of the battery.

3. Wrap a piece of wire around the side of the bulb. Place the other end of wire over the paper clip. Tape the end of wire to the cardboard without letting the wire touch the paper clip. When a magnet is placed near the paper clip, it will rise, touching the wire and turning on the light. (You can also use the Sneaky Ring to activate this magnetic switch.)

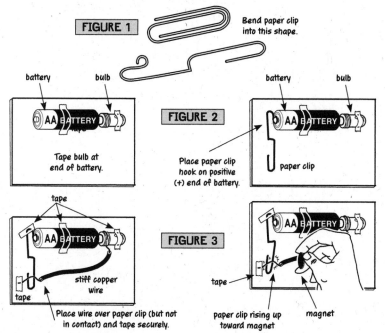

**FIGURE 1**

Bend paper clip into this shape.

battery    bulb

Tape bulb at end of battery.

**FIGURE 2**

battery    bulb

Place paper clip hook on positive (+) end of battery.

paper clip

tape

**FIGURE 3**

stiff copper wire

tape

Place wire over paper clip (but not in contact) and tape securely.

tape

paper clip rising up toward magnet

magnet

## MICRO-FOOSBALL GAME

1. Straighten four paper clips and bend them in the shape shown in **Figure 1**. Notice the **V** near the center, which acts as a paddle. Also notice the downward hook at one end. It prevents the paper clip paddle from sliding out of the paper clip holders. Wrap a small length of tape at the other end to act as a handle grip.

2. Carefully cut off the top flap and a square section at each end of the paper clip box. Slip four paper clips on each side.

3. Slide the four paper clip paddles into the paper clip loops on each side. Place the small candy ball on the game area, and you're ready to play.

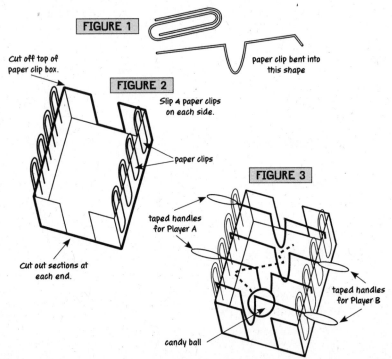

FIGURE 1

paper clip bent into this shape

Cut off top of paper clip box.

FIGURE 2

Slip 4 paper clips on each side.

paper clips

FIGURE 3

taped handles for Player A

Cut out sections at each end.

taped handles for Player B

candy ball

## TURN A PAPER CLIP INTO A HOUSE

Possibly the most amazing paper clip reuse of them all.

Kyle MacDonald, a Canadian blogger, bartered his way to a house starting with a single red paper clip (see **Figure 1**).

Here is the list of trades:

- ▶ Kyle traded the paper clip for a fish-shaped pen.
- ▶ He traded the pen for a hand-sculpted doorknob.
- ▶ He traded the doorknob for a camp stove.
- ▶ MacDonald traded the camp stove for a Honda generator.
- ▶ He traded the generator for an "instant party": an empty keg, an IOU for filling the keg with the beer of the holder's choice, and a neon beer sign.
- ▶ He traded the "instant party" to comedian Michel Barrette for a snowmobile.
- ▶ He traded the snowmobile for a two-person trip to Yahk, British Columbia.
- ▶ He traded the trip for a van.
- ▶ Kyle traded the van for a recording contract with the group Metal Works.
- ▶ He traded the recording contract for one year's rent in Phoenix, Arizona.
- ▶ Kyle traded the year's rent for one afternoon with musician Alice Cooper.
- ▶ He traded the time with Alice Cooper for a KISS motorized snow globe.
- ▶ Kyle traded the KISS motorized snow globe for a role in a Corbin Bernsen film.

▶ And last, Kyle traded the movie role for a two-story farmhouse in Kipling, Saskatchewan.

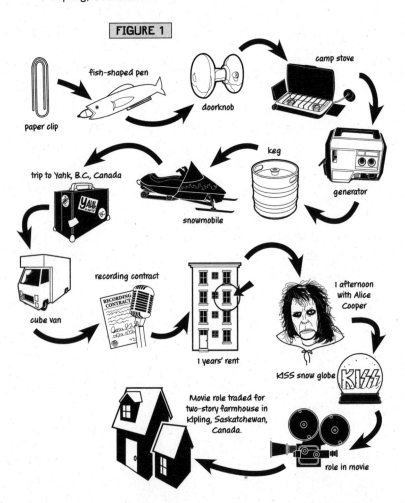

**FIGURE 1**

paper clip → fish-shaped pen → doorknob → camp stove → generator → keg → snowmobile → trip to Yahk, B.C., Canada → cube van → recording contract → 1 years' rent → 1 afternoon with Alice Cooper → KISS snow globe → role in movie → Movie role traded for two-story farmhouse in Kipling, Saskatchewan, Canada.

# SNEAKY FASHIONS

We all would like to stand out from the crowd without spending a fortune. You can! You just have to be *sneaky*. This section provides creative ways to modify your fashion accessories in ways that most people have not seen before.

First, you'll see how to make techno-fashion nails light up and animate figures on top; bet your friends have never seen that before! And, for a unique thrill, craft special techno-nails that display a secret message or image that your friends can see only with a sneaky invisible light!

Discover how to convert your bracelets into Sneaky Bangles that look stylish and conceal useful items. Learn how to make fun projects such as Sneaky Belts and Sneaky Headbands. For parties, craft a Sneaky Action T-Shirt or Sneaky 3D Party T-Shirt.

All the projects are tested safe and can be assembled quickly. You can put your new sneaky fashion skills to use making unique nails and accessories for family and friends, too.

# SNEAKY LIGHT-UP NAILS

Ever wanted to *really* stand out from the crowd? You could spend a small fortune on designer clothes or shoes or spring for expensive jewelry, but there is a sneakier way.

Using a micro LED light and a couple of tiny watch batteries, you can create unique illuminating nails. Simply adjusting the position of the top nail acts as an on/off switch, allowing your nails to outshine all others.

## What's Needed

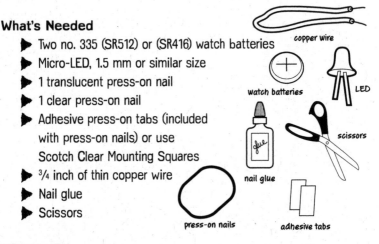

- ▶ Two no. 335 (SR512) or (SR416) watch batteries
- ▶ Micro-LED, 1.5 mm or similar size
- ▶ 1 translucent press-on nail
- ▶ 1 clear press-on nail
- ▶ Adhesive press-on tabs (included with press-on nails) or use Scotch Clear Mounting Squares
- ▶ ³/₄ inch of thin copper wire
- ▶ Nail glue
- ▶ Scissors

copper wire

watch batteries

LED

nail glue

scissors

press-on nails

adhesive tabs

## What to Do

Recent innovations in LED manufacturing have resulted in super-bright, low-power models in very tiny packages, tiny enough to be mounted between two press-on nails.

LEDs are electronic components that produce light when an electrical voltage, from a battery, is supplied to it. Unlike light bulbs,

LEDs produce almost no heat, but the wire leads must be connected to the battery in one direction only (with a small light, you could connect the leads in either direction to make it light up). Place the two batteries together and press the LEDs connector leads to each side. Ensure that the batteries are connected positive (+) to negative (−). If the LED does not light, reverse its leads by turning it around. See **Figure 1**.

Glue the micro LED on the center of the inside of the translucent nail as shown in **Figure 2**.

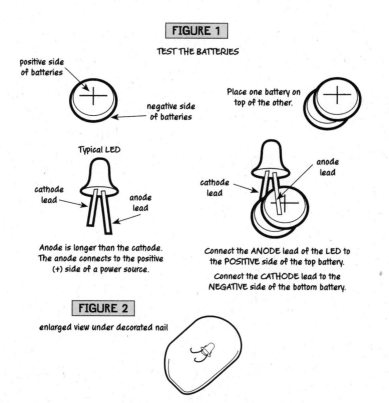

**FIGURE 1**

TEST THE BATTERIES

positive side of batteries

negative side of batteries

Place one battery on top of the other.

Typical LED

cathode lead

anode lead

Anode is longer than the cathode. The anode connects to the positive (+) side of a power source.

anode lead

cathode lead

Connect the ANODE lead of the LED to the POSITIVE side of the top battery.

Connect the CATHODE lead to the NEGATIVE side of the bottom battery.

**FIGURE 2**

enlarged view under decorated nail

On the lower clear nail, apply a small adhesive press-on tab strip at the center. Press the small piece of wire on the strip. See **Figure 3**.

Next, press the two batteries—one with the positive side up and the other with the positive side down—on the ends of the wire. Also apply adhesive strips at each end of the clear nail. See **Figure 4**. Besides bonding the nails together, the strips also raise the height of the ends of the lower nail, which has the raised batteries in the center, for an even fit.

The LED wire leads under the top nail and the batteries on the lower clear nail should be positioned so they contact the top of both batteries. See **Figure 5**.

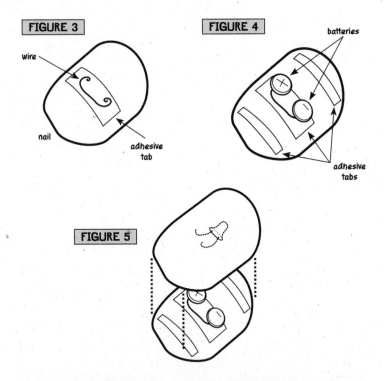

FIGURE 3

wire

nail

adhesive tab

FIGURE 4

batteries

adhesive tabs

FIGURE 5

When the parts are positioned properly, place the top nail on the bottom one and the LED should light. If it doesn't light, turn the top nail around and press it on. If necessary, bend the LED contacts upward so they make a solid connection. See **Figure 6**.

Apply press-on adhesive to your fingernail as shown in **Figure 7,** and press the lower nail on it. See **Figure 8.** Now you can press the top nail on the lower nail to make it light when desired as shown in **Figure 9**.

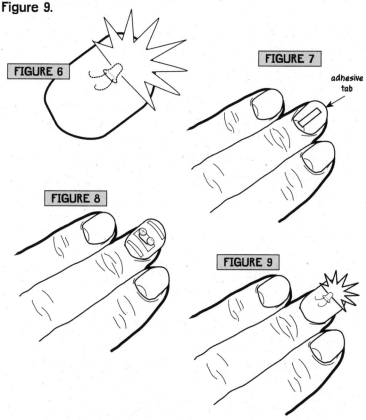

# SNEAKY ANIMATED NAILS

Advanced nail crafters produce amazing designs using large-sized nails
with 3-D figures mounted on top. It's now possible to animate these
items using a micro vibrating motor commonly used in cell phones.
Power is supplied by a single watch battery.

## What's Needed

- Micro Vibrating Motor
- No. 364 watch battery (or similar)
- Large press-on nail
- Adhesive press-on tabs (included
  with press-on nails) or use
  Scotch Clear Mounting Squares)
- Flexible ornament figure
- Small stick-on ornament
- Decorative stick-on decals
- Nail glue
- Scissors

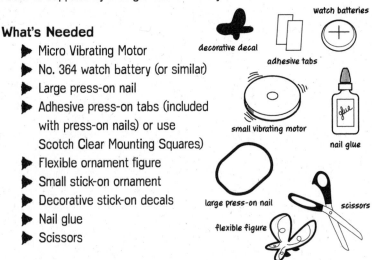

decorative decal

watch batteries

adhesive tabs

small vibrating motor

nail glue

large press-on nail

flexible figure

scissors

## What to Do

First test the vibrating motor by connecting its wire leads to both sides
of the watch battery as shown in **Figure 1**.

Decorate the press-on nail and the top and side of the vibrating
motor (if desired) with decorative stickers. See **Figure 2**.

Press on a rectangular strip of press-on adhesive near the center of
the nail as shown in **Figure 3**.

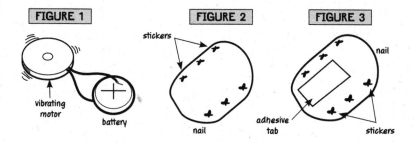

Place the vibrating motor near the front of the nail and press one of its wire leads firmly on the adhesive strip. **Note:** The vibrating motor must be able to move freely. **See Figure 4**.

Next, glue a small ornament, in this example a small plastic butterfly, to the remaining vibrating motor wire. Also glue the flexible figure to the top of the vibrating motor as shown in **Figure 5**. If desired, you can apply decorative stickers to conceal the vibrating motors wires.

When you press the small butterfly ornament, the top wire under it makes contact with the battery and activates the motor causing the butterfly on top to flutter! See **Figure 6**.

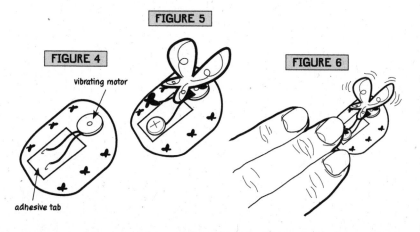

## GOING FURTHER

You can create more Sneaky Nail designs using the same parts and configuration used in the Sneaky Vibration Motor.

If you glue a small piece of stiff wire to the top of the vibrating motor and slip a small ornament with a center hole (the type used for making decorative beads), the item will spin. See **Figure 7.**

If necessary, bend the wire at an angle until the item, in this example a small soccer ball, spins as shown in **Figure 8.**

By substituting the Vibrating Motor for other microdevices, you can create even more Sneaky Nails, including:

TOP-OF-THE NAIL LIGHT: Attach the wires of a 1½-volt microbulb (used by model railroad hobbyists) to the battery to make your nail figures glow. See **Figure 9.**

SQUEALING NAIL: Mount a microbuzzer on top of the nail and conceal it under an animal figure to make a squealing nail design as shown in **Figure 10. Note:** You can place tape over the buzzer's top hole to alter its sound.

NAIL FAN: Similar to the Squealing Nail, mount a microfan to the top of a large nail and glue thin plastic strips on top. When you activate the fan, the streamers will fly upward. Or substitute a small square graphic drawn on thin plastic or coated paper and watch it rise when you activate the Sneaky Nail Fan as shown in **Figure 11.**

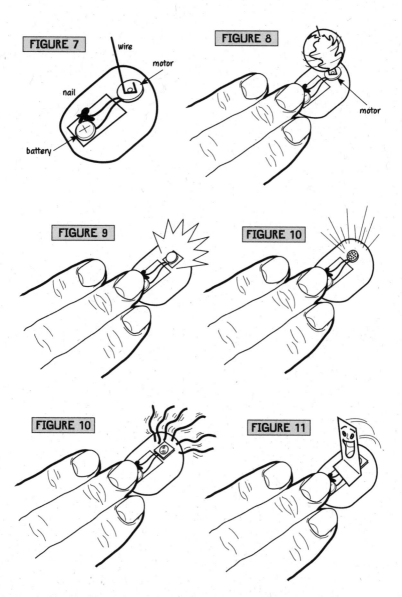

# INVISIBLE NAIL ART

Sometimes making a great impression doesn't mean showing off to everyone, just a certain someone. You can let your nails display one face to the public and a secret message to someone special with a sneaky two-way nail design. What's that, you say? It's a design where one part of the nail is visible and the other is not, unless you know the secret.

### What's Needed

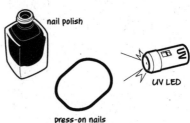

nail polish

UV LED

press-on nails

▶ Press-on nails
▶ Nail polish
▶ UV ink pen
▶ Micro-UV LED

## What to Do

First, decorate your press-on nails with standard polish and let them dry completely. See **Figure 1**.

Then apply a design on the nail with a UV applicator pen while shining the UV LED light on it, as shown in **Figure 2**.

Now you can show off your first visible nail design, and then show off the sneaky invisible creation when you expose it to UV light. See **Figures 3** and **4**.

**FIGURE 1**

polish

**FIGURE 2**

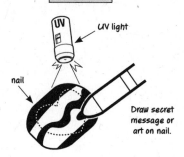

UV light

nail

Draw secret
message or
art on nail.

**FIGURE 3**

Affix press-on
nail to nail.

**FIGURE 4**

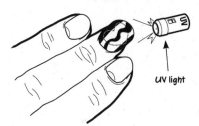

UV light

# SNEAKY BANGLE

If you thought bracelets and bangles were just for decoration, think again.

Your favorite bangle can provide more than just wrist decoration. By adding a few items, you can use it to verify your currency and provide protection for you and your purse.

## What's Needed

▶ Small extendable key fob
▶ Wide bangle or bracelet
▶ Velcro stick-on dots
▶ Mini round mirror(s)
▶ Micro-light(s)
▶ Optional: ornament
▶ Tiny, strong neodymium magnets

## What to Do

For this project you need to buy the smallest retractable key fob you can find that will fit inside the large bangle without being obtrusive; a loose-fitting bangle or bracelet is recommended.

First, press Velcro dots on the side of the bangle and the matching ones on a mirror and micro-light and press them securely in place. See **Figures 1** and **2**.

You can mount a small ornament on top of the micro-light's button to conceal it. See **Figure 3**.

Remove the key ring from the end of the key fob. Use Velcro dots to secure a small mirror on the end of the key fob chain. Attach Velcro dots or a small magnet to one side of the key fob. Attach matching pieces of Velcro dots or another magnet inside the bangle, and press securely in place. See **Figure 4**.

If you think your money is suspect, pull out the magnet and place it close to the edge of the currency to see whether it moves toward it. If it does, the bill is legitimate because real currency has iron particles in the ink.

Also, when you are walking down the street or at an ATM, you can pull out the mirror and see behind you. See **Figure 4**.

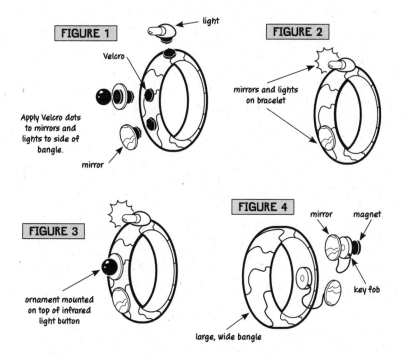

FIGURE 1

light

Velcro

Apply Velcro dots
to mirrors and
lights to side of
bangle.

mirror

FIGURE 2

mirrors and lights
on bracelet

FIGURE 3

ornament mounted
on top of infrared
light button

FIGURE 4

mirror    magnet

key fob

large, wide bangle

# SNEAKY BANGLE BAG

Combining accessories can be fun, and it gives you a unique look.
But sometimes if you mix unrelated accessories, like a tank top and a
bangle, you can create some sneaky utility, too.

Do you find yourself wishing you had a larger handbag to carry
items, and you just brought a small clutch purse? The Sneaky Bangle
Bag is the answer! It includes a compartment with a spare carryall bag
that unfolds at a moment's notice.

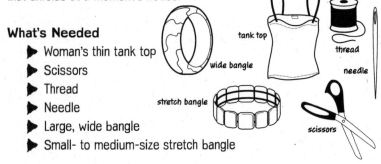

## What's Needed

▶ Woman's thin tank top
▶ Scissors
▶ Thread
▶ Needle
▶ Large, wide bangle
▶ Small- to medium-size stretch bangle

## What to Do

The carry bag will be made from a colorful, ultra-thin tank top. Be sure
to select one that has the same size neckline on the back and the front,
as shown in **Figure 1**.

Turn the tank top inside out and cut the bottom section to the
desired length for your custom bag. Sew the bottom shut and turn the
tank top right side out. See **Figure 2**.

Roll up the tank top, as shown in **Figure 3**.

Then, wrap the sneaky bag tightly around the wide bangle, as shown
in **Figure 4**.

Mount the stretch bangle over the larger bangle and bag. See **Figure 5**.

Now, you have multiplied your fashion accessory possibilities. You can wear each bangle separately or stretch the small one on top of the larger one. Also, the sneaky tank top bag can change the look of the smaller bangle. With the smaller bangle and tank top pair mounted on the larger bangle you've got a fifth look. And, when desired, you can unwrap the tote bag at a moment's notice. See **Figure 6**.

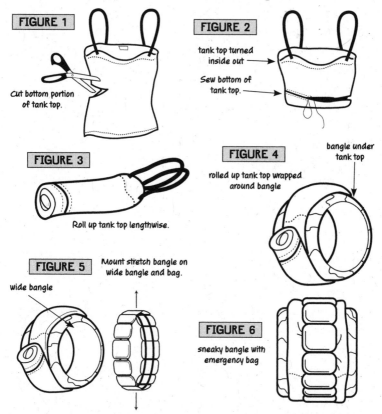

**FIGURE 1**

Cut bottom portion of tank top.

**FIGURE 2**

tank top turned inside out

Sew bottom of tank top.

**FIGURE 3**

Roll up tank top lengthwise.

**FIGURE 4**

rolled up tank top wrapped around bangle

bangle under tank top

**FIGURE 5**

Mount stretch bangle on wide bangle and bag.

wide bangle

**FIGURE 6**

sneaky bangle with emergency bag

# SNEAKY HEADBAND

Most women own many fashionable ornaments and have plenty of earrings and pendants that they don't wear often. You'll see how you can use them to decorate your favorite headband.

Conversely, you can use headband ornaments to change the look of your earrings and pendants.

## What's Needed

- Wide headband
- Glue
- Small, strong magnets
- Earrings and pendants with a flat surface
- Extra decorative ornaments
- Paper clips
- Glue or Velcro stick-ons (you can use either Velcro stick-ons or magnets for this project)

## What to Do

Work with earrings and pendants on a nearly flat surface to make it easy to mount the various ornaments.

Glue about six small magnets to the surface of the headband. See **Figure 1**.

Then, glue small, right-angle pieces of paper clips to the back of the ornaments and to the front of the earrings and pendants, as shown in **Figure 2**.

Mount the ornaments to the headband, earrings, and pendants as shown in **Figure 3**. If desired, you can also swap the ornaments from the headband to the earrings or the pendant.

## GOING FURTHER

You can perform this same versatile jewelry trick with other accessories, including your bracelets and belts.

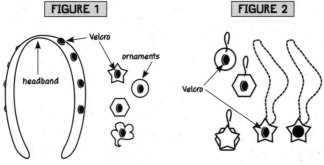

FIGURE 1

Apply Velcro on ornaments and on headband.

FIGURE 2

Apply Velcro on earrings and pendant.

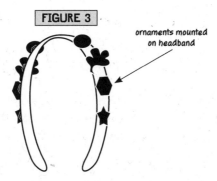

FIGURE 3

ornaments mounted on headband

# SNEAKY BELT

A great entrance is priceless, but we all know that high fashion can sometimes be painful. Arriving at a social event without some comfortable accessories can make you uncomfortable later on.

How would you like to appear at a social gathering with just a form-fitting dress, high heels, and a clutch bag, but leave the event in comfort and ready for travel? You just have to make our Sneaky Belt so you can change into more functional flat shoes, throw on a cap, and put on a top to help cover your arms and shoulders.

## What's Needed

- Woman's tank top
- Scissors
- Thread
- Needle
- Four elastic hair bands
- Wide belt
- Roll-up flat shoes (Dr. Scholl's or similar)
- Thin cap (or headband)
- Thin knit top

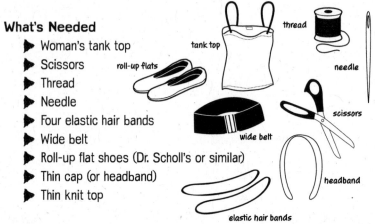

## What to Do

First make the tank top bag as shown in the Sneaky Bangle Bag project.

Next, cut the elastic hair bands into three long pairs and one short pair of straps and sew them into the back of the belt as shown in **Figure 1**. (Ensure that the flat shoes will be held in place before sewing on the elastic hair bands.)

Place the tank top bag, flat shoes, cap, and top in the loops as shown in **Figure 2**.

**Figure 3** shows a woman with a stylish appearance at an event.

After the event, when it's time to go outside, she can remove the cap, top, flat shoes, and bag from inside the belt to be more comfortable and be better protected from the elements. See **Figure 4**.

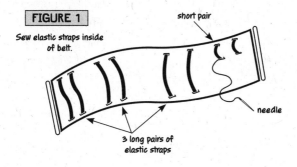

FIGURE 1

Sew elastic straps inside of belt.

short pair

needle

3 long pairs of elastic straps

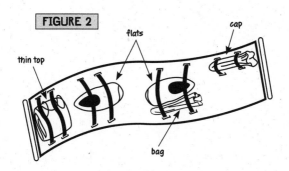

FIGURE 2

flats

cap

thin top

bag

FIGURE 3

FIGURE 4

Cap protects head.

Warm top protects shoulders and arms.

comfortable flats

Carry bag with shoes and clutch purse inside.

## GOING FURTHER

See how to store more sneaky accessories in your own Wild, Wild Vest in the Bonus Sneaky Projects section.

# SNEAKY ACTION T-SHIRT

Afraid of appearing out of style in your old T-shirts? Well, you won't anymore. Using items you already have, you can make a Sneaky Action T-Shirt that has detachable flying accessories that will intrigue your friends.

## What's Needed

- Scissors
- Cloth
- Pins
- Needle
- Strong nylon thread
- Optional: Velcro dots
- T-shirt or jacket
- Cardboard
- Tape

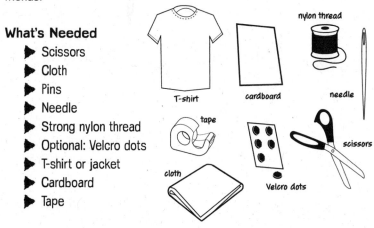

## What to Do

Cut cloth material (e.g., from a second T-shirt) into a 7-inch rectangular shape, fold over the outer edges, pin in place, and sew the edges, as shown in **Figures 1** and **2**. This will be mounted on the T-shirt as a "chest plate" to hold your action items.

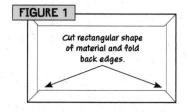

FIGURE 1

Cut rectangular shape of material and fold back edges.

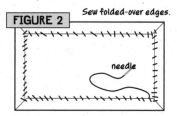

FIGURE 2

Sew folded-over edges.

needle

You can then sew the cloth chest plate onto the T-shirt. Or, if you prefer, you can use Velcro dots on the cloth to mount it on the T-shirt. See **Figure 3**.

Next, decorate the shirt with educational giveaway toys. Cut four pieces of cardboard into right angles that resemble small boomerangs, as shown in **Figure 4**.

Also, cut eight 2-inch squares of cardboard as shown in **Figure 5**. To make a Frisbee-like flying disc, fold the pieces of cardboard in half once and then fold one corner to the bottom. See **Figures 6a** and **6b**.

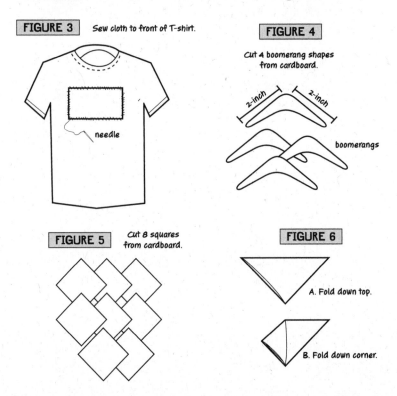

**FIGURE 3**  Sew cloth to front of T-shirt.

needle

**FIGURE 4**

Cut 4 boomerang shapes from cardboard.

2-inch  2-inch

boomerangs

**FIGURE 5**  Cut 8 squares from cardboard.

**FIGURE 6**

A. Fold down top.

B. Fold down corner.

Insert one piece into another until all eight pieces form a disc and tape them together. Fold down the outer edges. See **Figures 6c, 6d,** and **6e**.

Place the cardboard boomerangs on the front of the T-shirt's cloth cover and cut slits to hold them. Do the same with the cardboard flying disc, as shown in **Figure 7**.

Now slip the boomerangs and disc into the slits. Note that more than one item can fit per slit. Now you're ready to launch your sneaky boomerangs and flying disc from your Sneaky Action T-shirt. See **Figure 8**. Fly your sneaky aeronautical toys in public and show how they are made and flown.

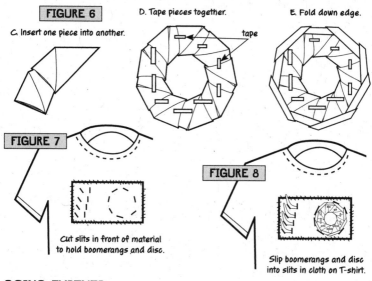

FIGURE 6

C. Insert one piece into another.

D. Tape pieces together.

tape

E. Fold down edge.

FIGURE 7

Cut slits in front of material to hold boomerangs and disc.

FIGURE 8

Slip boomerangs and disc into slits in cloth on T-shirt.

## GOING FURTHER

See more details on how to make and fly the Sneaky Mini Boomerang and Sneaky Flying Disc in the Bonus Sneaky Projects section.

# SNEAKY 3D PARTY T-SHIRT

Can't decide what to wear to a fun social event? Why not walk in with what appears to be a plain T-shirt and transform it into an attention-getting spectacle?

Using just two T-shirts and some other easily found items, you can make a Sneaky 3D Party T-Shirt to make a statement everywhere you go.

## What's Needed

- Two dark T-shirts
- Scissors
- Velcro dots
- Musical greeting card
- Large pop-up greeting card
- LED flasher pins

LED flasher pins

Happy Birthday musical greeting card

Velcro dots

two dark T-shirts

scissors

pop-up card

## What to Do

Cut a large rectangular piece from the front of T-shirt 1. See **Figure 1**.

Apply Velcro dots to T-shirt 2 and attach the cloth to the front of T-shirt 2. Leave the bottom half free to move, as shown in **Figure 2**.

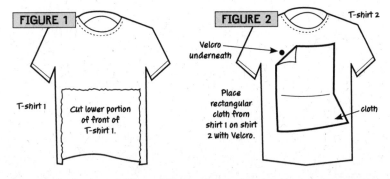

**FIGURE 1**

T-shirt 1

Cut lower portion of front of T-shirt 1.

**FIGURE 2**

T-shirt 2

Velcro underneath

Place rectangular cloth from shirt 1 on shirt 2 with Velcro.

cloth

Mount the musical greeting card on the cloth and then place the pop-up card on top of it with Velcro. Mount the LED flasher pins in each corner of the cloth. See **Figure 3**.

Now, when the cloth is folded up and held securely by the Velcro, the T-shirt looks normal. See **Figure 4**.

**Figure 5** shows how the Sneaky Party T-Shirt appears when you pull down the front cloth. The cards open, pop up, and play music. You can turn on the LED flashers manually.

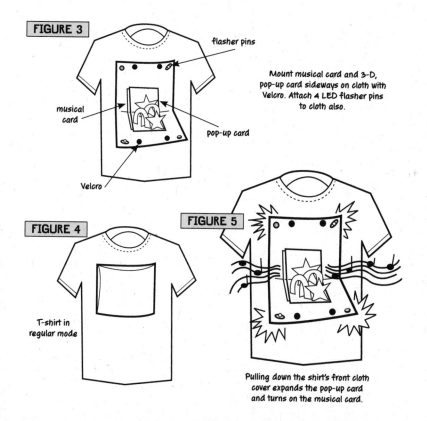

**FIGURE 3**

flasher pins

Mount musical card and 3-D, pop-up card sideways on cloth with Velcro. Attach 4 LED flasher pins to cloth also.

musical card

pop-up card

Velcro

**FIGURE 4**

T-shirt in regular mode

**FIGURE 5**

Pulling down the shirt's front cloth cover expands the pop-up card and turns on the musical card.

## GOING FURTHER

To make a Super Sneaky Party T-Shirt, obtain an inexpensive flat portable DVD player, as used on the back of automobile headrests for rear passenger viewing, to become a portable entertainment center. You can connect a portable battery pack and store it in your pocket or cargo pants. See **Figure 6**.

Add Internet access to the mix: Use Velcro stick-on dots to mount an inexpensive Wi-Fi–enabled tablet computer to the front of your shirt for online fun, as shown in **Figure 7**. For added fun, you can activate the tablet's camera function to provide onlookers with a digital mirror.

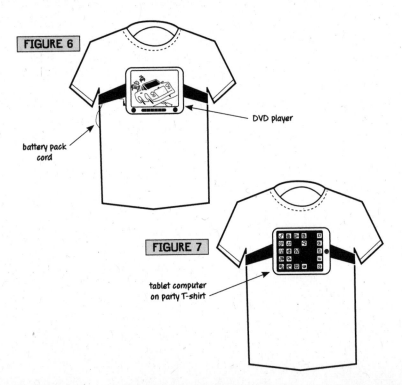

FIGURE 6

DVD player

battery pack cord

FIGURE 7

tablet computer on party T-shirt

# SNEAKY TOYS AND GAMES

Have you ever watched magicians levitate and wished you could do that, too? Or wished you could add to the challenge of beer pong by introducing a moving target? How about making a foosball game to play when you're on the road?

Did you know that an air cannon toy can be made with a grocery bag? Or that you can turn a musical greeting card into a room alarm? Adapt a toy radio-controlled car into a secret Nerf gun shooter? Well, you can, using just items found in any home.

In this section you'll also learn how to easily assemble and use a secret throwie signaler that only you or your friends can see (if they know the secret).

# LEVITATING GHOST

Magic! We've all done it. And often the sneakiest tricks involve
levitation. Seeing something float in midair defies our senses and is
mystifying. Here's a neat trick you can do that will astound onlookers.
With just a grocery bag, some wool, and a plastic pipe, you can create
a real levitating "ghost" that also demonstrates the magic of static
electricity.

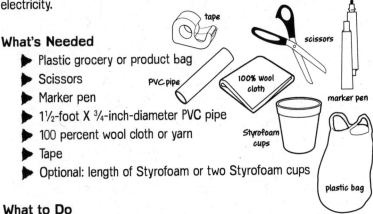

## What's Needed

▶ Plastic grocery or product bag
▶ Scissors
▶ Marker pen
▶ 1½-foot X ¾-inch-diameter PVC pipe
▶ 100 percent wool cloth or yarn
▶ Tape
▶ Optional: length of Styrofoam or two Styrofoam cups

## What to Do

Cut a rectangular section of plastic that is 1 inch wide and 3 inches
long. Cut the bottom two-thirds of the plastic into thin strips, like a
comb. Then draw little eyes on the top section, as shown in **Figure 1**.

Then rub the PVC pipe with the wool at least ten times in the same
direction. See **Figure 2**.

Hold the sneaky "ghost" and rub it across the wool at least ten
times, causing it to become negatively charged, as shown in **Figure 3**.

Raise the sneaky ghost high in the air over the pipe and let go. If
it sticks to your hand, shake it off vigorously until it floats free. Hold

the pipe away from your body and underneath the sneaky "ghost" and watch the ghost magically float up to 8 inches above the pipe. With a little practice you should be able to walk around with it by tilting the pipe in front of you. See **Figure 4**.

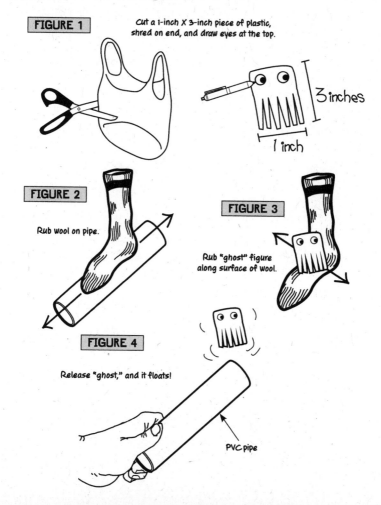

**FIGURE 1**

Cut a 1-inch X 3-inch piece of plastic, shred on end, and draw eyes at the top.

3 inches

1 inch

**FIGURE 2**

Rub wool on pipe.

**FIGURE 3**

Rub "ghost" figure along surface of wool.

**FIGURE 4**

Release "ghost," and it floats!

PVC pipe

## GOING FURTHER

Substitute other plastic items for the PVC pipe, such as a length of Styrofoam, as shown in **Figure 5**, or two Styrofoam cups taped together at the mouth, as shown in **Figure 6**.

And as a sister project, see how to make Sneaky Floating Photos in the Bonus Sneaky Projects section.

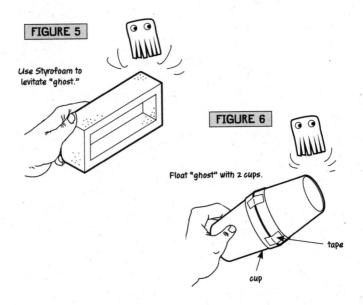

FIGURE 5

Use Styrofoam to levitate "ghost."

FIGURE 6

Float "ghost" with 2 cups.

tape

cup

# DIGITAL PICTURE FRAME APPLICATIONS

You have opportunities to meet new people when you walk down the street, wait in line at a store, or wait at a traffic light. Why not surprise them by sending them a sneaky electronic message to make an unforgettable impression?

The following project illustrates innovative ways to get more out of digital picture frames than just displaying pictures of friends and relatives.

### What's Needed

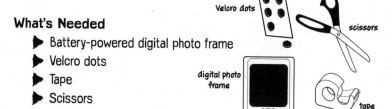

Velcro dots

scissors

- Battery-powered digital photo frame
- Velcro dots
- Tape
- Scissors

digital photo frame

tape

### What to Do

Add a digital frame to your favorite T-shirt
(after preparing it as shown in the
Sneaky 3D Party T-Shirt project),
as shown in **Figure 1**.

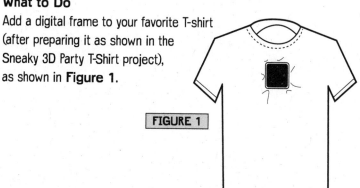

FIGURE 1

Load a few simple messages into a digital picture frame, mount it inside your top side window molding with Velcro stick-ons, and take advantage of stoplights and traffic jams to start up a conversation. See **Figure 2**.

Remove the ornament from a cheap ring and mount your digital frame on top of it with Velcro stick-ons. See **Figure 3**.

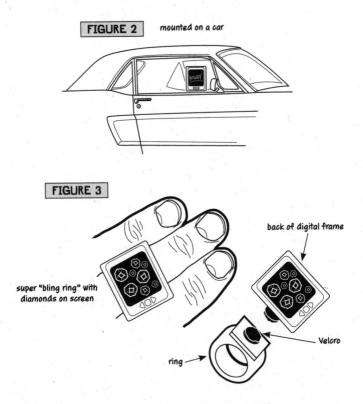

FIGURE 2   *mounted on a car*

FIGURE 3

*back of digital frame*

*super "bling ring" with diamonds on screen*

*Velcro*

*ring*

Perform the same technique with a wide cuff or belt, as shown in **Figures 4** and **5**. (Load the digital frame with test questions to help cram for a test.)

Digital picture frames can also be mounted on large sunglasses or earrings. See **Figures 6** and **7**.

Make a plush animal toy and insert a frame inside, loaded up with animal faces for fun. See **Figure 8**.

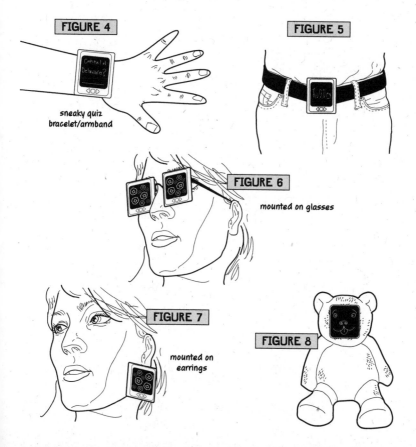

FIGURE 4

sneaky quiz
bracelet/armband

FIGURE 5

FIGURE 6

mounted on glasses

FIGURE 7

mounted on
earrings

FIGURE 8

# SNEAKY VIBRA-CUP

Playing beer pong can be fun, but if your aim is good, the challenge is gone. There is a sneaky way to increase the difficulty for advanced players: a randomly moving target! You'll see how with this sneaky project.

## What's Needed

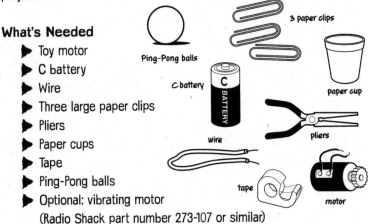

Ping-Pong balls

3 paper clips

C battery

paper cup

wire

pliers

tape

motor

- Toy motor
- C battery
- Wire
- Three large paper clips
- Pliers
- Paper cups
- Tape
- Ping-Pong balls
- Optional: vibrating motor
  (Radio Shack part number 273-107 or similar)

## What to Do

Test the toy motor with a single C battery by pressing its wires on both terminals. You can leave the connecting wires loose and turn on the motor by twisting them together.

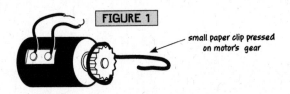

**FIGURE 1**

small paper clip pressed
on motor's gear

Next, bend and press a large paper clip around the gear of the motor and leave a looped section sticking up at the end. This will affect the balance of the motor and cause it to vibrate, as shown in **Figures 1** and **2**. Optionally, you can obtain small vibrating motors from discarded cell phones and pagers or purchase them from electronic supply stores.

Bend two paper clips into **C** shapes and tape them to the bottom edges of a cup. See **Figure 3**.

Then tape the motor and battery to the bottom of the cup. Turn the cup over, and it should vibrate across a flat surface, as shown in **Figure 4**. **Note:** By adjusting the length of the paper clip "feet" you can make the sneaky cup move around in a desired direction or pattern.

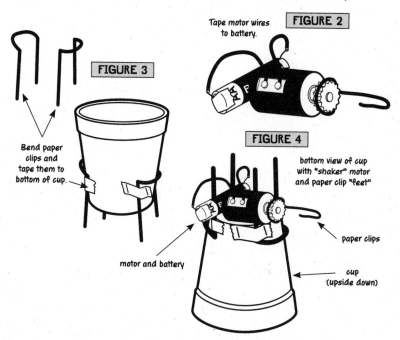

Tape motor wires to battery.

FIGURE 2

FIGURE 3

Bend paper clips and tape them to bottom of cup.

FIGURE 4

bottom view of cup with "shaker" motor and paper clip "feet"

paper clips

motor and battery

cup (upside down)

Now try to toss a Ping-Pong ball into the moving cup and challenge your friends to do the same at various distances. See **Figure 5**.

## GOING FURTHER

Add some drama by setting a predetermined time limit after each player starts the game.

See the next project for Sneakier Vibra-Cup Add-Ons.

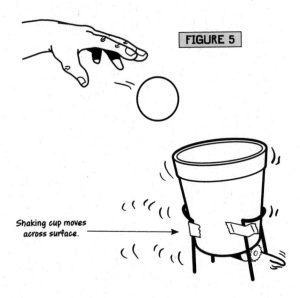

FIGURE 5

Shaking cup moves across surface.

# SNEAKIER VIBRA-CUP ADD-ONS

Soon your skills will increase, and even the Sneaky Vibra-Cup will not be as enjoyable as it once was. Add the following Sneakier Vibra-Cup Add-Ons to increase the scoring difficulty to bring the challenge back:

Have someone count down the time to score with a stopwatch or a digital watch or clock to add a level of suspense to the game.

To make it more difficult to score, place tape on the top sides of the mouth of the cup, as shown in **Figure 1**.

Or slip some rubber bands across the outer edges of the cup. See **Figure 2**.

You can also bend one end of a paper clip into a 45-degree angle and slip it onto the lip of the cup. See **Figure 3**. Add a couple of paper clips for a greater challenge, as shown in **Figure 4**.

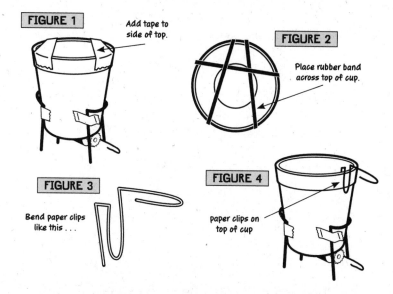

**FIGURE 1** Add tape to side of top.

**FIGURE 2** Place rubber band across top of cup.

**FIGURE 3** Bend paper clips like this . . .

**FIGURE 4** paper clips on top of cup

Use paper clips to add more "sidecar" cups to the Sneaky Vibra-Cup. See **Figure 5**.

Cut a hole at the bottom side of one or more cups to make it a challenge of not only tossing in a ball but keeping it in a cup until the time period expires. See **Figure 6**.

Add up to four cups mounted at 45-degree angles on the main Sneaky Vibra-Cup. Draw numbers on the sides for scoring purposes. Tossing the ball in an outer cup should produce a higher score than the center one. See **Figure 7**.

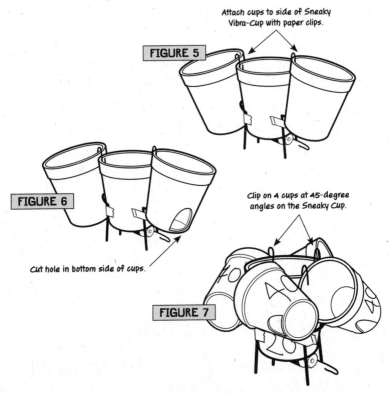

Attach cups to side of Sneaky Vibra-Cup with paper clips.

FIGURE 5

FIGURE 6

Cut hole in bottom side of cups.

Clip on 4 cups at 45-degree angles on the Sneaky Cup.

FIGURE 7

For added fun, lay out a game surface using spare poster board or flat corrugated cardboard from a large box. Fold up the outer edges so the vibra-cup stays inside the game area. Tape magnets to the board and draw scoring circles near them. Use pliers to break a small magnet into little bits and glue a few tiny magnet bits to the ball. Then start the game. If the ball rolls out of a cup with a hole in its side and makes it to a magnet, the player wins that score. See **Figure 8**.

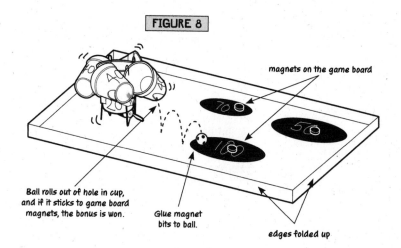

FIGURE 8

magnets on the game board

Ball rolls out of hole in cup, and if it sticks to game board magnets, the bonus is won.

Glue magnet bits to ball.

edges folded up

# VIBRATING FOOTBALL GAME

Using the same vibrating motor design shown in the Sneaky Vibra-Cup project, you can construct a Sneaky Vibrating Football (or Soccer) Game. Instead of attaching the vibra-motor to a cup, you'll mount it directly to the game surface, causing the figures on top to move around the playing field.

## What's Needed

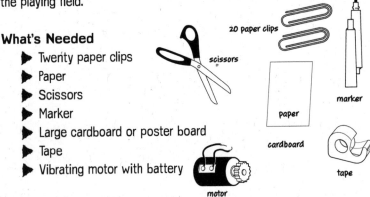

- Twenty paper clips
- Paper
- Scissors
- Marker
- Large cardboard or poster board
- Tape
- Vibrating motor with battery

20 paper clips

scissors

marker

paper

cardboard

tape

motor

## What to Do

For this project you will use a large piece of cardboard or poster board as a game board and mount the vibrating motor and battery on the board with tape.

Bend one end of a paper clip straight up, as shown in **Figure 1**.

Draw small pictures of your "players" on cut-out pieces of paper. See **Figure 2**.

Poke the straight end of the paper clip through the paper to make a game player. Repeat these steps to make as many players as needed for the game. See **Figure 3**. Designate a special player on each team as the ball carrier.

Bend the poster board so it has a slightly raised section in the middle. Tape the vibrating motor securely to the poster board, as shown in **Figure 4**. You can draw in lines on the field and a field goal section at each end.

The motor vibrates the board and the "players" on top. See **Figure 5**. Be sure to position some books or other objects at each end of the board so it doesn't move around.

To make the players move in a preferred direction, bend their paper clip supports as desired.

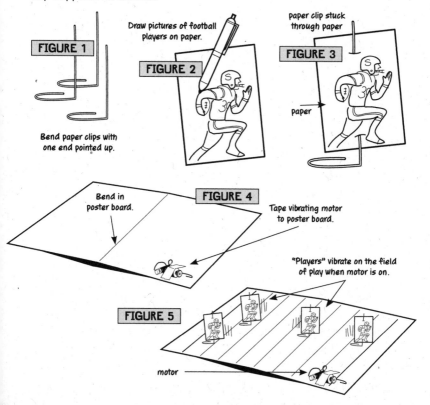

FIGURE 1

Bend paper clips with one end pointed up.

Draw pictures of football players on paper.

FIGURE 2

paper clip stuck through paper

FIGURE 3

paper

Bend in poster board.

FIGURE 4

Tape vibrating motor to poster board.

"Players" vibrate on the field of play when motor is on.

FIGURE 5

motor

# SNEAKY PORTABLE ALARM

Wouldn't it be great to have a cheap, lightweight, portable alarm for travel and regular household use without having to assemble it? If you have a musical greeting card, you've already got a position sensor, power source, sound circuitry, and speaker all in a cheap package ready to be put in use. You just have to make some sneaky modifications.

## What's Needed

▶ Musical greeting card
▶ Thread
▶ Tape
▶ Paper clip
▶ Velcro stick-on

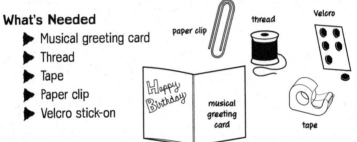

## What to Do

**Figure 1** shows a typical musical greeting card. The left side has a fold-over cover that conceals the parts underneath.

Open the covered side of the card that conceals the speaker and circuit board and you'll also see a plastic pull tab on the left that acts as a switch. When you open the card, the pull tab, which separates two metal pieces, slides to the left and connects the switch, activating the circuit. See **Figure 2**. First, carefully separate the pull tab from the left side of the card so the card can be opened fully without activating the alarm.

Next, tape thread at the top of the door you want to protect, as shown in **Figure 3**.

Tape the paper clip to the top of the wall next to the door and lead the thread through the paper clip and over and under the tape on the door. See **Figure 4**.

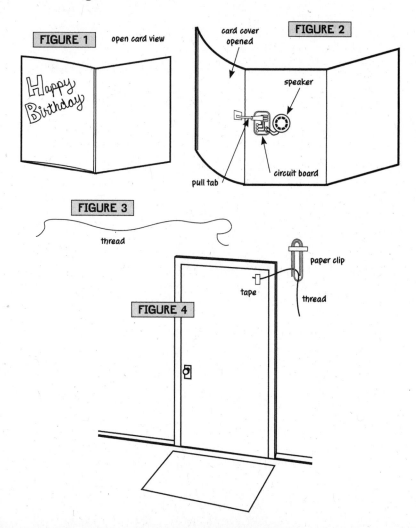

FIGURE 1    open card view

*Happy Birthday*

FIGURE 2    card cover opened

speaker

circuit board

pull tab

FIGURE 3

thread

FIGURE 4

paper clip

tape

thread

Lead the thread back to the card and tape it to the pull tab, as shown in **Figure 5**.

Tape the card on the wall near the door (or sitting on a table or shelf nearby) and adjust the tension on the thread properly. Ensure that the musical card does not move.

Now when the door opens, it pulls the thread, which pulls on the card's switch and activates the sneaky alarm, as shown in **Figure 6**.

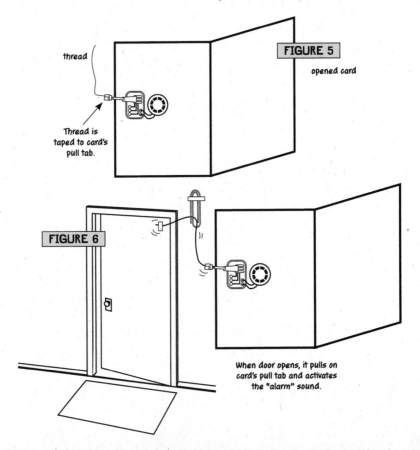

thread

Thread is taped to card's pull tab.

**FIGURE 5**

opened card

**FIGURE 6**

When door opens, it pulls on card's pull tab and activates the "alarm" sound.

# SNEAKY FOOSBALL GAME

Foosball (or paddle ball) games are a lot of fun, but unless you're at a bar or club, you usually cannot enjoy playing them. That's about to change.

With items you already have, you can make a home-style foosball game for you and your friends in no time.

## What's Needed

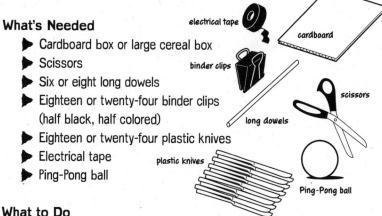

electrical tape

cardboard

binder clips

scissors

long dowels

plastic knives

Ping-Pong ball

▶ Cardboard box or large cereal box
▶ Scissors
▶ Six or eight long dowels
▶ Eighteen or twenty-four binder clips (half black, half colored)
▶ Eighteen or twenty-four plastic knives
▶ Electrical tape
▶ Ping-Pong ball

## What to Do

The size of the box is not critical. Just be sure to obtain dowels (long, round pieces of wood) that are at least 3 inches wider than the box.

Use a sturdy rectangular box for the frame. Cut off its top flaps and make holes for the goal area at each end, as shown in **Figure 1**.

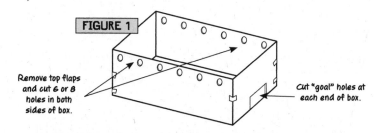

FIGURE 1

Remove top flaps and cut 6 or 8 holes in both sides of box.

Cut "goal" holes at each end of box.

Cut six to eight holes along each side for the dowels to slide in, as shown in **Figure 2**.

Next, clamp a binder clip on a dowel and place a plastic knife between the clamp and the dowel. Check the length so it can swing freely, and cut off the excess length as needed. See **Figure 3**.

Add more binder clips and knife "paddles" to the remaining dowels. Each dowel should have two or three paddles, depending on the width of the box used. Every other dowel should have black binder clips attached to it so players will know which ones to control during play. Wrap black electrical tape around the ends of every other dowel on each side of the foosball game with the black binder clips to make it easy for players to make quick choices during game play.

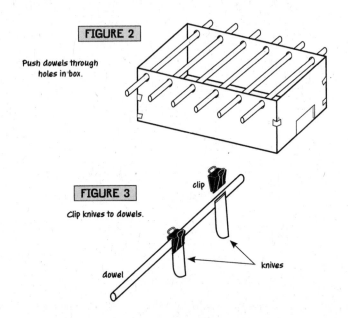

**FIGURE 2**

Push dowels through holes in box.

**FIGURE 3**

Clip knives to dowels.

clip

dowel

knives

Cut one hole on the left side to launch the Ping-Pong ball, as shown in **Figure 4**. Now you're ready to play.

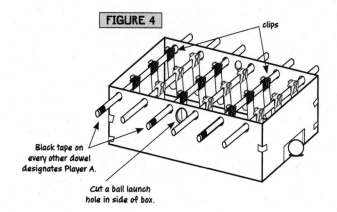

FIGURE 4

clips

Black tape on
every other dowel
designates Player A.

Cut a ball launch
hole in side of box.

# SNEAKY FOOSBALL GAME 2

What if you want to make a personal foosball game but don't have access to wooden dowels or a large box? Fear not: You can still make an even more compact, portable Sneaky Foosball Game with items that almost everyone has.

This design is smaller and can even be transported in a backpack for fun on car trips, while camping, anywhere!

### What's Needed

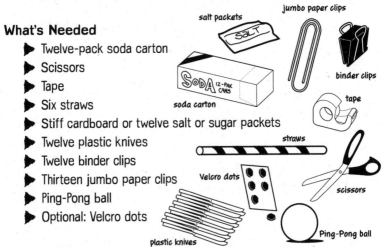

salt packets

jumbo paper clips

soda carton

binder clips

tape

straws

Velcro dots

scissors

plastic knives

Ping-Pong ball

- ▶ Twelve-pack soda carton
- ▶ Scissors
- ▶ Tape
- ▶ Six straws
- ▶ Stiff cardboard or twelve salt or sugar packets
- ▶ Twelve plastic knives
- ▶ Twelve binder clips
- ▶ Thirteen jumbo paper clips
- ▶ Ping-Pong ball
- ▶ Optional: Velcro dots

### What to Do

Cut off the top of the carton and tape the flaps securely on the side of the box. Also, cut open "end zone" scoring flaps for the ball to exit at each end of the carton, as shown in **Figure 1**.

The straws need something inside them to make them rigid. You can roll up salt or sugar packets found at fast-food outlets or wads of thin cardboard. See **Figure 2**.

Cut the top two inches off of twelve plastic knives, as shown in **Figure 3**.

Clamp two knives on each straw with the binder clips, as shown in **Figure 4**. The straw–knife units will act as game paddles.

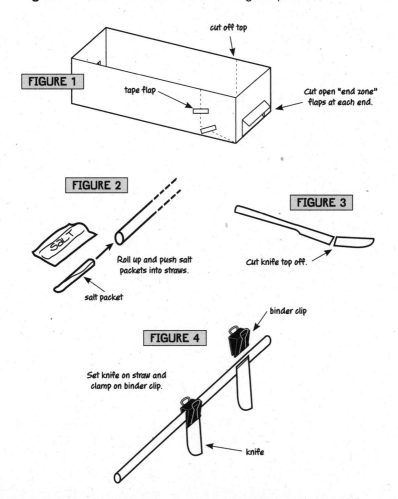

**FIGURE 1**

cut off top

tape flap

Cut open "end zone" flaps at each end.

**FIGURE 2**

SALT

salt packet

Roll up and push salt packets into straws.

**FIGURE 3**

Cut knife top off.

**FIGURE 4**

binder clip

Set knife on straw and clamp on binder clip.

knife

Next, slip six jumbo paper clips on each side of the carton to act as paddle guides. See **Figure 5**.

Bend a paper clip into the clip-on shape shown in **Figure 6** to act as a ball holder. Place it at the center of the game and push the ball off its holder to start game play.

Slip the game paddles into the paper clip guides, and you're ready to play! See **Figure 7**.

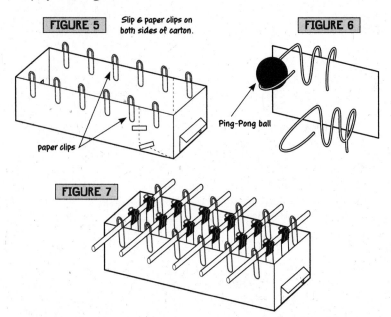

FIGURE 5    Slip 6 paper clips on both sides of carton.

paper clips

FIGURE 6    Ping-Pong ball

FIGURE 7

## GOING FURTHER

For easy disassembly and reassembly, you can carefully separate the sides of the carton and stick Velcro dots on the corners. With the carton neatly folded you can store the carton, straws, clips, and ball in a backpack for transportable play.

# MINI-FOOSBALL GAME

You can make an even more compact Sneaky Foosball game with items that are usually thrown away. With just a single three-bag microwave popcorn box and some paper clips and straws, you can quickly put together your own Mini-Foosball Game.

## What's Needed

▶ Three-bag popcorn box
▶ Scissors
▶ Tape
▶ Eight jumbo paper clips
▶ Four straws
▶ Gumball or jawbreaker candy

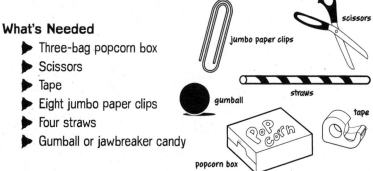

jumbo paper clips

scissors

gumball

straws

tape

popcorn box

## What to Do

Open the popcorn box and fold the cardboard flat, as shown in **Figure 1.** Cut away the front of the box, but don't throw it away; you'll use it to make game paddles.

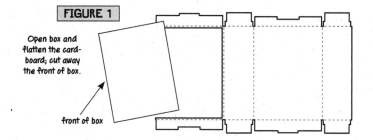

FIGURE 1

Open box and flatten the cardboard; cut away the front of box.

front of box

Fold the cardboard back into a box shape and tape the ends together. Also cut away holes at each end for the game ball to exit. See **Figure 2**.

Next, slide eight paper clips onto the top sides of the box, four per side, as shown in **Figure 3**.

Then cut six paddle pieces, square on one end and oval at the other, from the cardboard you saved earlier. Each piece should be approximately ¾ inch wide and 2 inches long. See **Figure 4**.

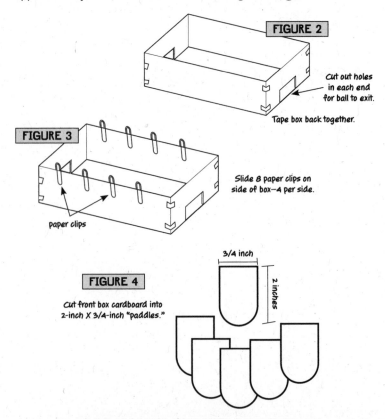

**FIGURE 2**

Cut out holes in each end for ball to exit.

Tape box back together.

**FIGURE 3**

Slide 8 paper clips on side of box—4 per side.

paper clips

**FIGURE 4**

Cut front box cardboard into 2-inch X 3/4-inch "paddles."

3/4 inch

2 inches

Tape two paddles near the center of each straw, as shown in **Figure 5**. At each end, tape one paddle to a straw to act as a goalie.

Last, slip the straws into the paper clip holders, and you're set. Place the "ball" into the game board area and start the action. See **Figure 6**.

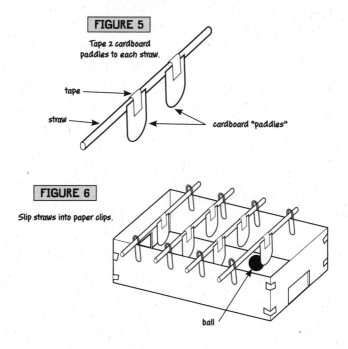

FIGURE 5

Tape 2 cardboard
paddles to each straw.

tape

straw

cardboard "paddles"

FIGURE 6

Slip straws into paper clips.

ball

# SNEAKY INVISIBLE SIGNALER

Every household has old remote controls from discarded appliances and devices that are useless. Or are they? Those remote controls have a special LED inside that emits invisible light, which you can put to sneaky use.

With a spare remote control and a tiny watch battery, you can easily make a miniature secret signaler to sneakily communicate with your friends.

## What's Needed

- Remote control from a discarded TV, DVD player, or other appliance
- Pliers
- 3-volt watch battery (CR2016 or similar)
- Camera phone or digital camera
- Micro-LED
- Tape

pliers

tape

watch battery

camera phone

micro-LED

## What to Do

LEDs are electronic components that produce light when an electrical voltage, from a battery for instance, is supplied. Unlike light bulbs, LEDs produce almost no heat, but the wire leads must be connected to the battery in one direction only (with a small light, you could connect the leads in either direction to make it light up).

Consumer remote controls use invisible infrared light to control devices via a receiver inside the appliance. A little-known fact is that camera phone and digital camera viewfinders can detect infrared light.

You can use this sneaky fact to make a secret communicator, which can be used at public gatherings.

First, remove the batteries and screws from the remote control case. Use pliers to carefully bend and cut the infrared LED leads from the circuit board.

You can test the infrared LED with a 3-volt watch battery and your camera. Position the wire leads on the infrared LED so they will be touching each side of the battery. While holding the infrared LED, activate the camera and look through the viewfinder to see whether it lights. When viewed from a camera phone or digital camera viewfinder, the LED's normally invisible light will appear white. If not, reverse the leads. Note the position and tape one lead to the battery. See **Figure 1**.

Next, open the case of the micro-LED, as shown in **Figure 2**.

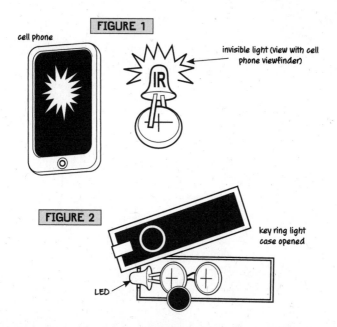

**FIGURE 1**

cell phone

invisible light (view with cell phone viewfinder)

IR

**FIGURE 2**

key ring light case opened

LED

Carefully remove the LED inside. See **Figure 3**.

Place the infrared LED in the case in the same position as the LED. Test it with the cell phone as before and, if necessary, reverse its position until it lights. See **Figure 4**.

Replace the micro-LED case screws, and now you have a sneaky LED signaler. Create a few sneaky flash codes that only your friends know, or use Morse code signals (search online for a Morse code list). You can communicate at parties, in the mall, and at other social gatherings without anyone else detecting your secret messages. See **Figure 5**.

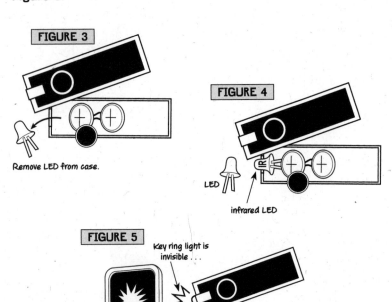

FIGURE 3

Remove LED from case.

FIGURE 4

LED

infrared LED

FIGURE 5

Key ring light is invisible . . .

. . . unless viewed through a cell phone screen (or digital camera).

# SNEAKY AIR CANNON

Kids love shooter video games, but what if you could make a sneaky toy that shoots a powerful stream of air to knock over real-life objects?

Believe it or not, you can do it using everyday items found in almost every home.

## What's Needed

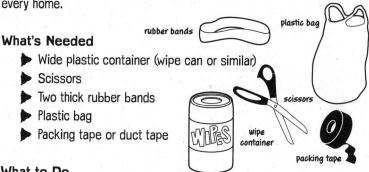

- Wide plastic container (wipe can or similar)
- Scissors
- Two thick rubber bands
- Plastic bag
- Packing tape or duct tape

## What to Do

Cut a 2-inch round hole in the bottom of the container, as shown in **Figure 1**.

Next, cut the top and inner flaps from the top of the container. See **Figure 2a**.

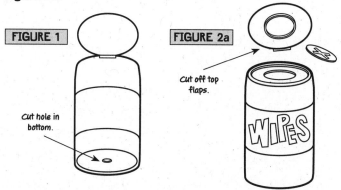

Next, cut a rubber band into a single long piece. Tape its ends to each side of the inside of the can about 2 inches from the top of the container. Also, tape the center of the rubber band to the inside of the bag. See **Figures 2b, 2c,** and **2d.**

Cut a portion of plastic from the bag and cover the top of the container with it. Leave the plastic loose on the top section, as shown in **Figure 3.**

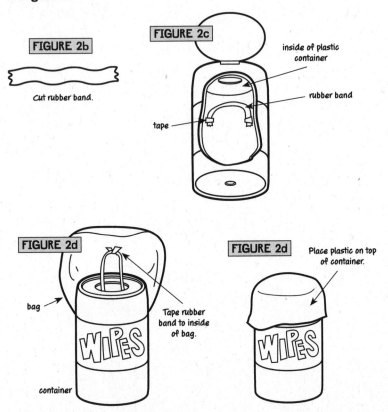

FIGURE 2b

Cut rubber band.

FIGURE 2c

inside of plastic container

rubber band

tape

FIGURE 2d

bag

Tape rubber band to inside of bag.

container

FIGURE 2d

Place plastic on top of container.

WIPES

WIPES

Tape the plastic around the side of the container securely. See **Figure 3**.

Then position the second rubber band in the center of the plastic on the top of the container, as shown in **Figure 4**.

Tape the second rubber band securely to the top center of the plastic bag, as shown in **Figure 5**.

To operate the Sneaky Air Cannon, hold the container on its side with one hand and pull back on the rubber band with your other hand, as shown in **Figure 6**. This will draw air into the container and into the bag.

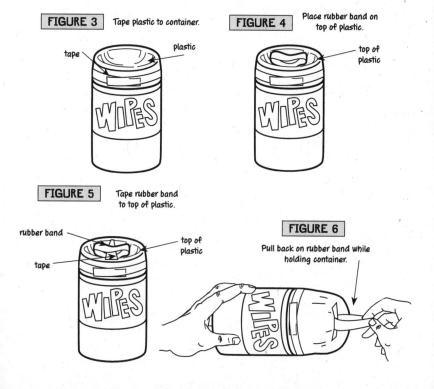

**FIGURE 3**   Tape plastic to container.

tape   plastic

**FIGURE 4**   Place rubber band on top of plastic.

top of plastic

**FIGURE 5**   Tape rubber band to top of plastic.

rubber band   top of plastic

tape

**FIGURE 6**

Pull back on rubber band while holding container.

Release the rubber band. When it snaps back it pulls the plastic bag into the container with a high-pressure force, and the air inside it is pushed out through the hole in the container's bottom. See **Figure 7**.

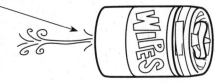

**FIGURE 7**

Releasing rubber band forces high-pressure air out of bottom hole.

# SNEAKY THROWIES

Throwies are little magnetically mounted LED lights that urban pranksters throw up to land on pipes and lampposts. Want to leave a signal for sneaky scavenger hunts that is in full view yet only detectable by you and your friends?

You can with a Sneaky Invisible Throwie.

## What's Needed

- Flashing LEDs
- Infrared LEDs
- 3-volt batteries (CR2032 or similar)
- Electrical tape
- Neodymium magnets

flashing LEDs

infrared LEDs

electrical tape

3-volt batteries

magnets

## What to Do

Connect the ends of an LED to a button-cell battery, and it will light. If not, reverse the connections, as shown in **Figure 1**.

Wrap the LED–battery pair with tape. See **Figure 2**.

Tape a small magnet to the outer portion of tape and wrap tape around it, as shown in **Figure 3**.

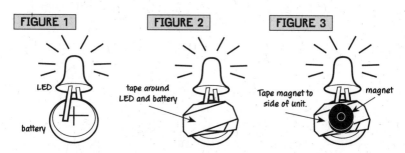

FIGURE 1
LED
battery

FIGURE 2
tape around
LED and battery

FIGURE 3
Tape magnet to
side of unit.
magnet

Now look for an area that has a high metallic object such as a pole, drain, or bracket and throw your Sneaky Throwie at it until the magnet sticks to it. See **Figure 4**. The light should stay lit for about two weeks with a strong battery.

For a sneakier signal, substitute an infrared LED for the standard type, and then only you and your friends can detect the invisible light by looking though the viewfinder of a digital camera or camera phone. See **Figure 5**.

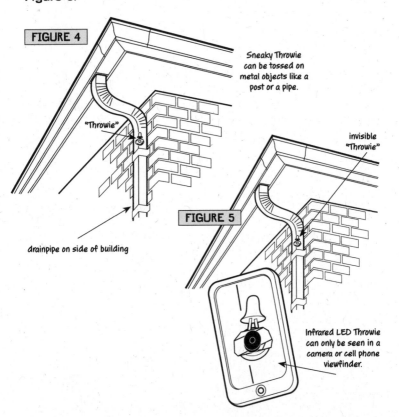

FIGURE 4

Sneaky Throwie
can be tossed on
metal objects like a
post or a pipe.

"Throwie"

invisible
"Throwie"

FIGURE 5

drainpipe on side of building

Infrared LED Throwie
can only be seen in a
camera or cell phone
viewfinder.

# REMOTE-CONTROLLED NERF SHOOTER

Nerf guns and Ping-Pong ball shooter toys are fun, but your friends can usually see you coming and avoid getting tagged. Want to amaze your friends (and pets)?

Build this remote-controlled Nerf gun shooter you can make with everyday things to bring a sneakier dimension to game play.

## What's Needed

▶ Toy Nerf gun or pop gun
▶ Pliers
▶ Screwdriver
▶ Toy radio-controlled truck
▶ Packing or duct tape

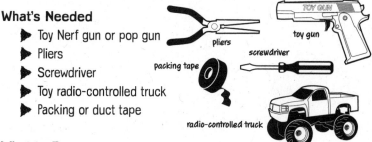

pliers

toy gun

screwdriver

packing tape

radio-controlled truck

## What to Do

Select a simple-looking Nerf gun or Ping-Pong ball shooter that does not have anything in front of the trigger guard. (Some toy-gun designs have storage compartments mounted in front of the trigger guard.)

Break off the trigger guard with the pliers, as shown in **Figure 1**.

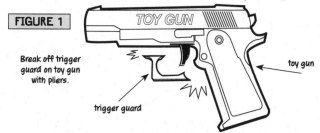

**FIGURE 1**

Break off trigger guard on toy gun with pliers.

toy gun

trigger guard

Next, using a screwdriver, remove the front wheel assembly cover from the toy truck, and remove the wheels. See **Figure 2**.

You should see a steering shaft that moves back and forth when you switch on the remote control LEFT and RIGHT knob. See **Figure 3**.

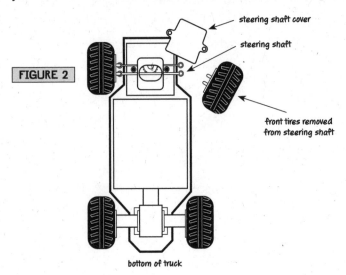

steering shaft cover

steering shaft

**FIGURE 2**

front tires removed
from steering shaft

bottom of truck

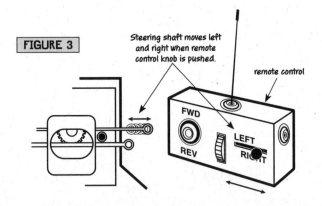

**FIGURE 3**

Steering shaft moves left
and right when remote
control knob is pushed.

remote control

FWD

REV

LEFT

RIGHT

Position the toy gun next to the car's steering shaft. Activate the shaft with the remote to ensure that it will push the trigger. Once the proper position is found, mount and tape it very securely. Set up the truck and toy gun pair so when activated they will not move when the foam Nerf "bullet" is fired. See **Figure 4**. Now you can hide the remote-controlled toy shooter and secretly activate it by pressing the RIGHT button on the remote control. See **Figure 5**.

For more sneaky action, remove the control board from the plastic case, substitute small watch batteries for the larger types, and hide the device in your pocket or jacket.

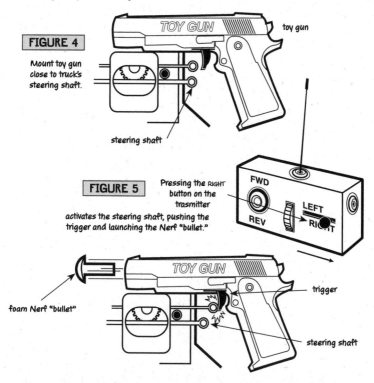

FIGURE 4

Mount toy gun
close to truck's
steering shaft.

toy gun

steering shaft

FIGURE 5

Pressing the RIGHT
button on the
trasmitter
activates the steering shaft, pushing the
trigger and launching the Nerf "bullet."

FWD

LEFT

REV

RIGHT

foam Nerf "bullet"

trigger

steering shaft

## GOING FURTHER

Make the remote-controlled car's steering shaft activate rubber band–powered toys and shooters.

Remotely turn on a musical greeting card with the remote-controlled car's steering shaft.

Remove the transmitter from its case, substitute smaller button cell batteries, and hide it inside your belt buckle or jacket sleeve.

For more radio-controlled car applications, see Sneaky Radio-Controlled Car Projects in the Bonus Sneaky Projects section.

# SNEAKY SCIENCE PROJECTS

If you're curious about science and technology, you're about to discover plenty of sneaky projects that you can put together in your home.

Did you know that you can power devices with your plants? Or make a battery with a pencil? If you've ever made a coin-battery science project, you learned you have to use two different types of coins, such as a penny and a nickel. But is there a sneaky way to make one with just a penny? Do you know the sneaky secret that lies inside a 9-volt battery?

Or, if you don't have a solar cell for a science project, what other electronic component generates electricity when it's exposed to light? You soon will know all these answers and more.

Want to know how to control other devices with your motion detector? How to adapt a mind control toy without opening up its case? It's all right here.

All these sneaky science projects are ready for quick construction with items already at your disposal.

This section also includes an introduction to going beyond everyday things to develop your own sneaky projects with special sound, touch, light, and other sensors. You'll learn how to use microcontroller kits to design programmable devices with your computer, including high-tech fashion projects.

Let's get started!

# POWER DEVICES
# WITH YOUR PLANTS

Many people know that you can make a battery using fruits and two different metals to power low-powered devices. Wait, you didn't know that? First see "Make Batteries from Everyday Things" in the "Bonus Sneaky Projects" section for more information.

You can go further and access more sneaky power for science experiments. You can tap electrical power by taking advantage of the latent energy inside plants. By connecting a series of plants together with common wire, you can demonstrate how this works and even power a light.

## What's Needed

- ▶ Cup
- ▶ Small plants
- ▶ Water
- ▶ Copper wire
- ▶ Paper clips
- ▶ Voltmeter
- ▶ Low-voltage LED or light or 1½-volt-powered clock
- ▶ Styrofoam egg carton

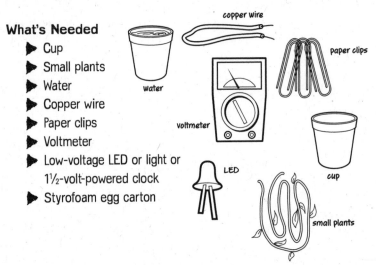

## What to Do

To see plant power in action, straighten a paper clip and stick it into the cup. Then strip the ends from a 6-inch length of copper wire and place it into the cup, as shown in **Figure 1**.

Next, place some plants into the cup and fill it with water. Stir the cup with a straightened paper clip. Then connect the voltmeter leads to the ends of the wire and paper clip. In a few minutes the readings should be near ½ volt. Add some salt to the solution to see whether the voltage rises. See **Figure 2**.

To power a device such as an LED, you'll need to connect multiple plant batteries in series with each other; you can make a sneaky multiplant holder out of an egg carton! Place three small plants in each compartment of the egg carton. Simply place a paper clip, copper wire, and plant with water inside each compartment. Then connect each section to another with connecting wire. Be sure to connect the wire from the paper clip to the copper wire of each compartment. Then test the collective voltage level. See **Figure 3**.

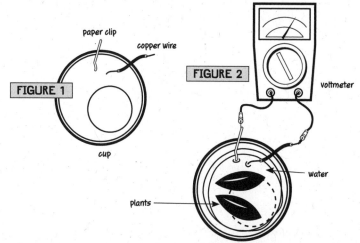

If necessary, add more plants to more egg carton compartments and connect them in series until you can achieve 2 volts or more. Then, connect the wires to a low-voltage light, which operates on about 1½ volts, or to another device, such as an LCD clock, which operates with only one AA or watch battery. You can also connect a small LED to the plant battery array.

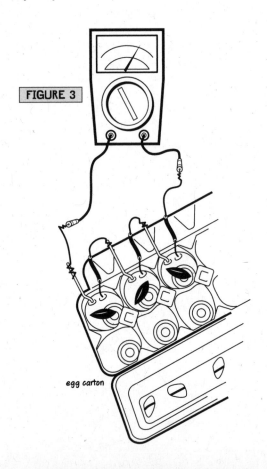

FIGURE 3

egg carton

# PENCIL BATTERY

When manipulated by a skilled mathematician, writer, or artist, a pencil can do some amazing things.

Were you aware that an ordinary pencil can also provide electrical power? It can, and you'll see a sneaky adaptation that will make an interesting science project.

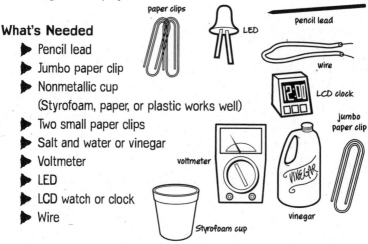

## What's Needed
- Pencil lead
- Jumbo paper clip
- Nonmetallic cup
  (Styrofoam, paper, or plastic works well)
- Two small paper clips
- Salt and water or vinegar
- Voltmeter
- LED
- LCD watch or clock
- Wire

## What to Do
Remove the lead from a pencil (or use lead from a mechanical pencil).
See **Figure 1**.

**FIGURE 1**

Crack open pencil and remove lead.

Next, straighten a jumbo paper clip and place it with the pencil lead into the cup as shown in **Figure 2**.

Carefully clip on the small paper clips to the tips of the lead and large paper clip and pour in the salt water (or vinegar). See **Figure 3**.

Connect the voltmeter leads to the two paper clips to see how much power, in volts, the sneaky Pencil Battery produces, as shown in **Figure 4**.

## GOING FURTHER

Make more pencil batteries and connect them in series by connecting wires from the lead to the large paper clip and so forth until you have enough power to activate an LED. See **Figure 5**.

You should also be able to power an LCD watch or clock with your Pencil Battery cells.

You can also experiment with different proportions of the ingredients in your liquid solution.

An array of pencil batteries can be connected together in series to provide more power to energize a clock or LEDs.

For another sneaky power trick, see the Sneaky Electrical Generator in the Bonus Sneaky Projects section.

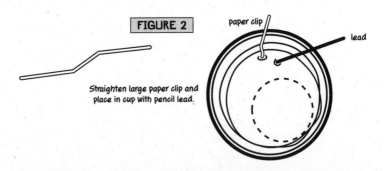

FIGURE 2

Straighten large paper clip and place in cup with pencil lead.

paper clip

lead

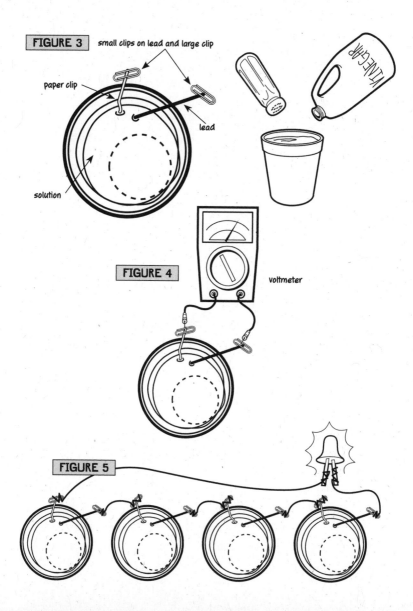

**FIGURE 3**

small clips on lead and large clip

paper clip

lead

solution

**FIGURE 4**

voltmeter

**FIGURE 5**

VINEGAR

# SNEAKY PENNY BATTERY

A sneaky battery can be made with two different metals, such as a penny and a nickel, with a moist paper towel soaked in lemon juice in between.

But what if you don't have a lot of nickels or quarters? Well, you're in luck. There is a sneaky way to make a battery just using a single type of coin.

## What's Needed

▶ Ten pennies
▶ Sandpaper
▶ Tissue or paper towels
▶ Salt water or vinegar
▶ Wire
▶ LED
▶ Plastic pill bottle
▶ Optional: voltmeter

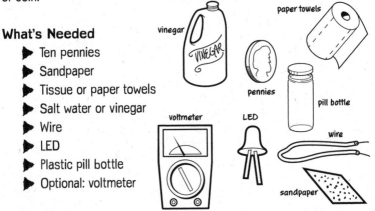

## What to Do

A penny isn't made of what it used to be. Pennies made before 1982 were a mixture of 95 percent copper and 5 percent zinc, but post-1982 pennies have a 97.5 percent zinc core with only a 2.5 percent copper plating.

You can take advantage of this bimetal combination and make a battery cell from just a single coin type. (Normally, you need to use two different coins and place a liquid solution between them to generate electrical power.)

First, sand or grind off the copper coating from just one side of ten post-1982 pennies, as shown in **Figure 1**.

Then dip nine small wads of tissue in a salt water solution or vinegar and place them between the ten pennies in a stack. Be sure to stack the pennies in the same direction, with the copper side on the bottom and the zinc side on the top. See **Figure 2**.

You can use a voltmeter to check the power of this stack of battery cells, as shown in **Figure 3**.

Once you have enough coins stacked to produce about 2 volts of power, connect wires to the ends of an LED and touch them to the top and bottom of the coin battery stack. If the LED does not light, reverse the wire connections. See **Figure 4**.

You can slip the penny battery inside a pill container tube for easy use, as shown in **Figure 5**.

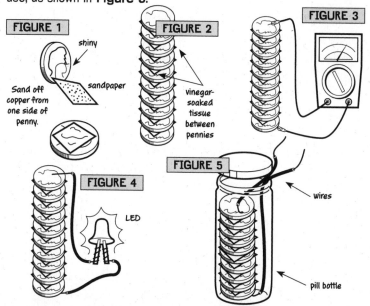

FIGURE 1
shiny
Sand off copper from one side of penny.
sandpaper

FIGURE 2
vinegar-soaked tissue between pennies

FIGURE 3

FIGURE 4
LED

FIGURE 5
wires
pill bottle

# SNEAKY 9-VOLT BATTERY TRICK

Are you desperate for some AA or AAA batteries, but can't find any? If you can locate a good 9-volt battery and some pliers, you're in luck!

This sneaky trick will show you how to turn a single 9-volt battery into six cells that can substitute for AAA or AA batteries in a pinch.

### What's Needed

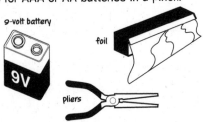

- ▶ 9-volt battery
- ▶ Pliers
- ▶ Optional: aluminum foil

### What to Do

Carefully pry open the case of the 9-volt battery, and you'll find six cylinders that are actually 1.4-volt AAAA batteries. See **Figure 1**. The batteries are connected together with thin metal strips, as shown in **Figure 2**. Carefully remove the strips at the bottom of the batteries with pliers to make proper contact.

In a pinch these batteries can fit into a AAA battery compartment easily enough. To make a secure fit, stretch out the battery compartment's spring (or metal tab) or add a piece of rolled-up aluminum foil. Optionally, the batteries can fit into a AA battery compartment. To see how the battery sizes compare, see **Figure 3**.

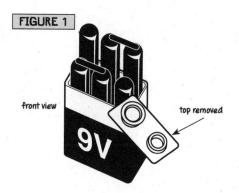

**FIGURE 1**

front view

top removed

9V

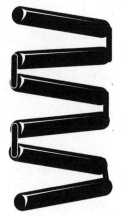

**FIGURE 2**

Six 1.4-volt AAAA cells are inside
the 9-volt case.

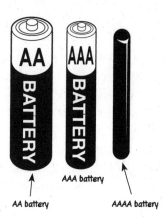

**FIGURE 3**

AA battery

AAA battery

AAAA battery

# SNEAKY DIODE POWER TRICK

A diode is an electronic component that conducts electricity in only one direction, like a one-way valve. Many types of diodes are designed for special purposes, such as detecting radio signals or producing illumination (LEDs, including infrared and ultraviolet LEDs). Some diodes even act as light sensors.

Sneaky fact: Almost all diodes are sensitive to light. In fact, LEDs can be used to not only emit light but also detect it. And some diodes can generate a small electrical current when exposed to light, just like solar cells. You can produce a nice science demonstration with the following technique.

## What's Needed

▶ Various diodes

▶ Pliers

▶ Voltmeter

▶ Bright lamp or LCD clock

LED

zener diode

signal diode

pliers

LCD clock

voltmeter

## What to Do

You can purchase diodes at electronic parts supply stores, or extract them from older electronic devices. Check the printed circuit boards for diodes and remove them with needle-nose pliers. Diodes have two connector leads and usually a dark band on one end (not to be confused with resistors, which appear similar but have many multicolored bands on them). See **Figure 1** for diode examples.

Connect your voltmeter to the connector leads of a typical diode and set it to the lowest voltage range on its scale. The reading should be zero, as shown in **Figure 2**.

Some signal diodes are very small and have a clear case. Other diodes, such as a rectifying diode, have a black or gray case. If you carefully squeeze this type of case with pliers, the case will break away to reveal the small diode inside, which is light-sensitive. See **Figure 3**.

Once you have located a few diodes that produce voltage levels under a bright lamp (see **Figures 4** and **5**), you may be able to connect them in series and produce more voltage, possibly enough to run a small LCD clock that needs only a single 1½-volt battery.

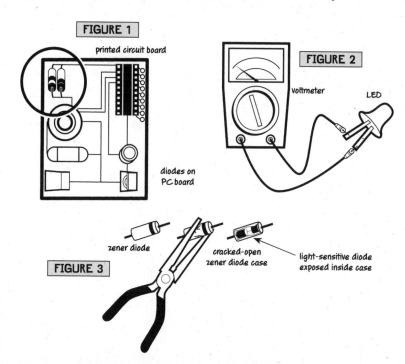

FIGURE 1

printed circuit board

diodes on PC board

FIGURE 2

voltmeter

LED

zener diode

cracked-open zener diode case

light-sensitive diode exposed inside case

FIGURE 3

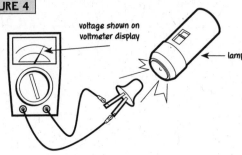

voltage shown on
voltmeter display

lamp

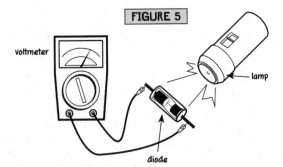

voltmeter

lamp

diode

# TOY MOTION-DETECTOR ADAPTATIONS

Many inexpensive, stand-alone security alarms and toys include motion-detector circuits. Wouldn't it be great if you could use them for more than setting off their own light or buzzer? These sneaky motion sensors can be easily transformed in just a few sneaky steps.

Once you see how easy it is to adapt these magical sensors, you'll be thinking of plenty of novel applications of your own.

## What's Needed

- Toy motion detector
- Pliers
- LED or battery-powered toy
- Electrical tape
- Scissors

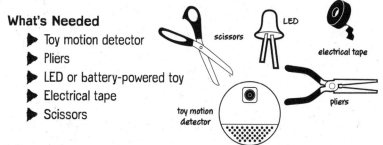

scissors

LED

electrical tape

pliers

toy motion detector

## What to Do

Most motion-detector toys activate a light or an alarm buzzer when something approaches them. The detector circuit inside supplies electrical power to a small buzzer, but you can tap this energy to activate other devices of your choosing.

**Note:** Be sure to test your detector for proper operation before performing this modification. Sensitivity can vary from one model to another.

Remove the case cover of the motion detector and place it aside. Look for the buzzer or LED indicator light, if included, and remove it from the case. See **Figure 1**.

Cut its two connecting wires and strip the insulation off the ends. Note how many batteries are in the motion detector and add up the voltage supplied. See **Figure 2**. Example: If the unit uses two 1½-volt AA batteries, then they can supply a maximum of 3 volts to the buzzer (although it could be a lower amount).

You can connect the two wires to another toy's battery connector, without its batteries installed, and activate the toy when your hand comes near the motion detector, as seen in **Figure 3**.

Optionally, a small relay, a device that isolates two electrical circuits from another, can be added to your motion-detector projects for more versatility. See **Figure 4**.

## GOING FURTHER

Attach an infrared LED for a secret visual indicator that only you can see by looking through your camera phone or digital camera viewfinder.

For more toy motion-detector adaptation ideas, see Sneaky Toy Modifications in the Bonus Sneaky Projects section.

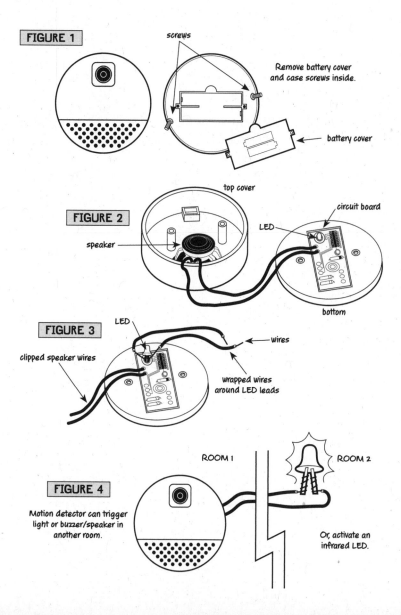

**FIGURE 1**

screws

Remove battery cover and case screws inside.

battery cover

**FIGURE 2**

top cover

circuit board

LED

speaker

bottom

**FIGURE 3**

LED

wires

clipped speaker wires

wrapped wires around LED leads

**FIGURE 4**

ROOM 1

ROOM 2

Motion detector can trigger light or buzzer/speaker in another room.

Or, activate an infrared LED.

# MIND-CONTROLLED TOY MODIFICATION

There are fantastic toys on the market that allow you to control devices with infrared light, radio signals, and even your mind!

If they have been sitting neglected in your closet, you're not taking full advantage of the sneaky resources you have at your disposal.

If you want to extend their applications beyond game play, this project will show you how without disassembling the case and voiding the warranty.

## What's Needed

▶ Mind-controlled toy, such as the Mind Flex or Force Trainer
▶ Jumbo paper clips
▶ Paper clips
▶ Cardboard
▶ Tape
▶ LED
▶ Wire
▶ 3-volt battery (CR2032 or similar)

## What to Do

**Figure 1** shows a typical mind-controlled toy that consists of a headset and a game device that blows air to levitate a ball. The more you concentrate, the higher you can levitate the ball. If you don't want to disassemble the case, there is a sneaky way to make this work for you.

The object is to make the air rushing upward activate a see-saw type of switch that will activate a simple circuit.

To start, bend a paper clip into the shape shown in **Figure 2**. Half the paper clip is flat on the cardboard and the other half is bent upward into an **S** loop.

Straighten a second paper clip and then bend it into a loop around the first one. Tape a square piece of cardboard on the shorter end to the left, as shown in **Figures 3a** and **3b**.

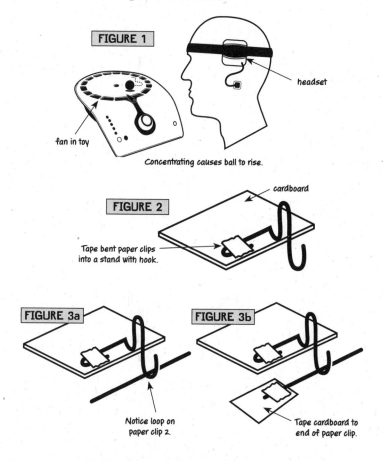

FIGURE 1

headset

fan in toy

Concentrating causes ball to rise.

FIGURE 2

cardboard

Tape bent paper clips into a stand with hook.

FIGURE 3a

Notice loop on paper clip 2.

FIGURE 3b

Tape cardboard to end of paper clip.

Straighten a third paper clip and tape it so it is positioned under the second paper clip. Also connect a 3-volt battery and LED to paper clips 1 and 3. See **Figure 4**.

With this arrangement you've created a see-saw tilt switch. Ensure that the balance is set so a slight rush of air will move the paper clip. When the air from the mind control game rushes upward, it pushes the cardboard side up and tilts down the other end of the paper clip. When it does, it completes the circuit and turns on the LED, as shown in **Figure 5**.

## GOING FURTHER

Want another sneaky way to use the mind control toy to activate other devices? Simply position a motion detector close to it so it detects the movement of the rising ball. With a little patience, obtain an arrangement where your thoughts elevate the ball, and the motion detector senses the change of position and activates the device of your choosing.

See the next section for more sneaky mind control device adaptation ideas.

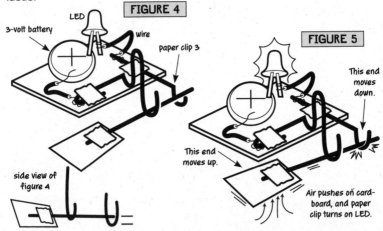

**FIGURE 4**

LED

3-volt battery

wire

paper clip 3

side view of figure 4

**FIGURE 5**

This end moves down.

This end moves up.

Air pushes on cardboard, and paper clip turns on LED.

# SNEAKY USES
# *BEYOND* EVERYDAY THINGS

This book presents a variety of ways to reuse everyday objects and adapt toys and devices to extend their applications. We hope you were intrigued with the ideas shown here, but you can go much further with your creative designs and increase your knowledge of technology by using additional low-cost sensors, science kits, and easy-to-use microcontroller kits on the market.

Purchase and experiment with science and chemistry kits. Play with robot construction kits and electronic project labs to learn more about mechanical models, transistors, integrated circuits, and relays to safely control more powerful devices. See **Figure 1**.

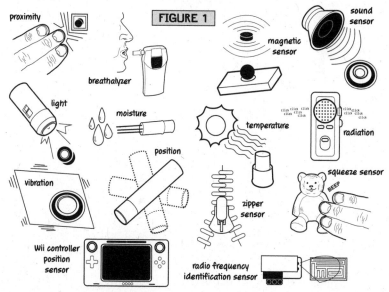

FIGURE 1

proximity

breathalyzer

sound sensor

magnetic sensor

light

moisture

temperature

click click click
click click
click

click click
click click
click

radiation

position

vibration

zipper sensor

squeeze sensor

BEEP

Wii controller position sensor

radio frequency identification sensor

Investigate microcontroller kits so you can make programmable smart projects. The Arduino is an inexpensive and popular standard that allows for a cheap and simple way to control sensors, lights, motors, and much more.

For example, you can use input signals from a variety of sensors, including proximity, breathalyzer, magnetic, sound, light, moisture, temperature, radiation, vibration, position and movement (including the position of a zipper or the pressure used to squeeze a plush toy), the motion from a Wii game controller's accelerometer, or your own position (detected from a radio frequency tag or a Global Positioning System chip), and precisely control an LED, solenoid or motor, toy car, robot or helicopter, door lock, and much, much more.

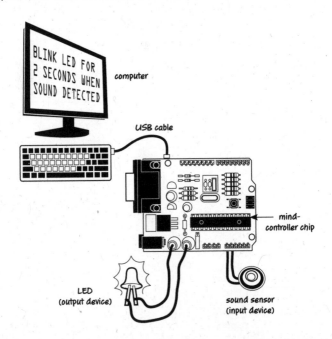

You can program what occurs and how. You simply connect your input and output devices to the Arduino board, connect it to your computer, apply your desired actions, disconnect from the computer, and you're done. Many museums incorporate Arduino interactive art displays that detect and track visitors and provide information accordingly.

Imagine the sensor-to-output device mashups you can make!

You can start with a simple project such as activating an LED when a sound is detected. Then move your way up to interfacing a microcontroller board to toy remote controls and create your own mind-controlled toy helicopter. Example sensors and applications are shown in **Figure 2**.

**FIGURE 2**

The user's thoughts are received by the mind-control
receiver connected to an Arduino microcontroller board,
which controls the helicopter's remote control.

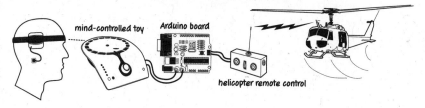

Investigate and experiment with micro-controller kits such as Phidgets, LilyPad, Lego Mindstorms, and BASIC Stamp 2. The LilyPad is designed for fashion-oriented tech projects, allowing you to use textile sensors to produce custom LED light shows on your clothes! See **Figures 3** and **4**.

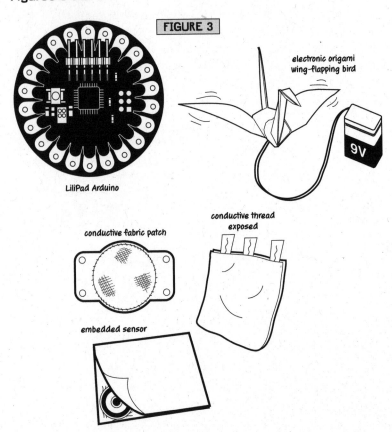

FIGURE 3

LiliPad Arduino

electronic origami wing-flapping bird

9V

conductive fabric patch

conductive thread exposed

embedded sensor

FIGURE 4

LiliPad Arduino kit

electronic clothing
creations

Combine origami creations with shape-shifting "muscle wire" to flap bird wings.

Make an LED suit that lights up your own custom light show displays.

An Arduino microcontroller project can be used to water plants when they're thirsty, turn on grow lights, and alert you with a Tweet or text message if the temperature drops.

Use a motion detector to trigger a radio-controlled transmitter, which activates a Nerf shooter. Or set up voice-activated devices that control just about any gadget or appliance you can imagine.

▶ Add piezo-speakers, sound sensors, timers, conductive thread and fabrics, textile sensors, and surface-mounted LEDs to your projects to make dynamic fashion creations.

▶ Investigate the LilyPad Arduino toolkit that enables people to build and program soft, flexible, fabric-based fashion designs. A set of sewable electronic modules enables people to blend textile craft and custom-programmed electrical engineering.

▶ Design interactive wallpaper that can be programmed to monitor its environment, control lighting and sound, and generally serve as a beautiful and unobtrusive way to enrich environments with computation. You could run your hand across this wallpaper to turn on a lamp, play music, or send a message to a friend. The wallpaper is flat, constructed entirely from paper and paint, and can be paired with a paper computing kit whose pieces serve as sensors, lamps, network interfaces, and interactive decorations.

# BONUS SNEAKY PROJECTS

# MAKE BATTERIES FROM EVERYDAY THINGS

No one can dispute the usefulness of electricity. But what do you do if you're in a remote area without AC power or batteries? Make sneaky batteries, of course!

In this project, you'll learn how to use fruits, vegetable juices, paper clips, and coins to generate electricity.

## What's Needed

▶ Lemon or other fruit
▶ Nail
▶ Heavy copper wire
▶ Paper clip or twist tie
▶ Water
▶ Salt
▶ Paper towel
▶ Pennies and nickels
▶ Plate

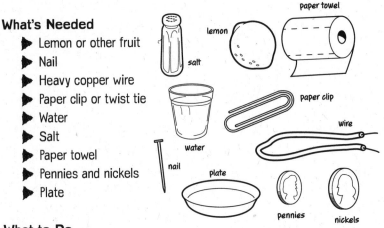

## What to Do

## The Fruit Battery

Insert a nail or paper clip into a lemon. Then stick a piece of heavy copper wire into the lemon. Make sure that the wire is close to, but does not touch, the nail (See **Figure 1**). The nail has become the battery's negative electrode and the copper wire is the positive electrode. The lemon juice, which is acidic, acts as the electrolyte. You can use other item pairs besides a paper clip and copper wire, as long as they are made of different metals.

The lemon battery will supply about one-fourth to one-third of a volt of electricity. To use a sneaky battery as the battery to power a small electrical device, like an LED light, you must connect a few of them in series, as shown in **Figure 2**.

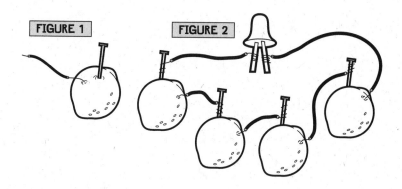

FIGURE 1     FIGURE 2

## The Coin Battery

With the fruit battery, you stuck the metal into the fruit. You can also make a battery by placing a chemical solution between two coins.

Dissolve 2 tablespoons of salt in a glass of water. This is the electrolyte you will place between two dissimilar metal coins.

Now moisten a piece of paper towel or tissue in the salt water. Put a nickel on a plate and put a small piece of the wet, absorbent paper on the nickel. Then place a penny on top of the paper (see **Figure 3**).

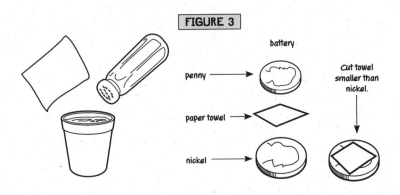

**FIGURE 3**

battery

penny →

paper towel →

nickel →

Cut towel smaller than nickel.

In order for the homemade battery to do useful work, you must make a series of them stacked up, as seen in **Figure 4.** Be sure the paper separators do not touch one another.

The more pairs of coins you add, the higher the voltage output will be. One coin pair should produce about one-third of a volt. With six pairs stacked up, you should be able to power a small flashlight bulb, LED, or other device when the regular batteries have failed. See **Figure 5.** *Power will last up to two hours.*

Once you know how to make sneaky batteries, you'll never again be totally out of power sources.

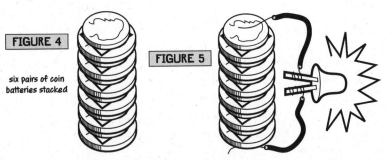

**FIGURE 4**

six pairs of coin batteries stacked

**FIGURE 5**

# SNEAKY RADIO-CONTROLLED CAR PROJECTS

Radio-controlled cars have many sneaky adaptation possibilities that can increase their usefulness. This project uses the inexpensive single-function type of radio-controlled toy car; this model will travel forward continuously, once its on/off switch is placed in the on position, until you actuate the remote control button, causing it to back up and turn. When you release the control, the vehicle goes forward in a straight line again.

The instructions and illustrations that follow will show you how to modify the transmitter to a more compact size, to use it as an alarm trigger. You'll also see how to modify the receiver to activate other devices, such as lights and buzzers.

## What's Needed

▶ Radio-controlled car
▶ Three 3-volt watch batteries
  (or fewer, depending on transmitter)
▶ LEDs
▶ Buzzer
▶ Tape
▶ Wire
▶ Rubber band
▶ Playing card
▶ Strong thin thread

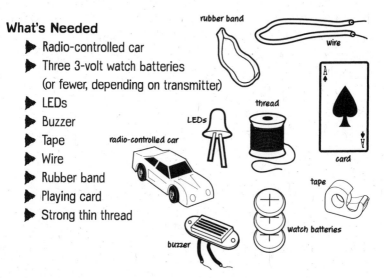

## What to Do

HOW A RADIO-CONTROLLED CAR WORKS. Pressing the transmitter button closes an electrical switch, which turns on the transmitter. This sends electromagnetic waves through the air that are detected by the radio receiver in the vehicle. The receiver detects the radio signal from the transmitter and reverses the electrical polarity (direction of current flow) of the power applied to the motor. This causes it to run in the reverse direction.

ADAPTING THE CAR'S TRANSMITTER. The first sneaky adaptation to the transmitter is to make it as small as possible, for concealment inside other objects or clothing.

Since the transmitter is always in the off mode until its activator button is pressed, it can operate using tiny long-life watch batteries.

If the transmitter uses one AA or AAA battery, it can be replaced by one small watch battery with the same voltage output. **Note:** Each AA or AAA battery supplies 1½ volts of power.

If the transmitter operates on two AA or AAA batteries, you can substitute either two 1½-volt watch batteries or a single 3-volt watch battery. If a 9-volt battery was in use, you will need to use three 3-volt watch batteries. When stacking batteries, place the positive side of one battery against the negative side of the other.

**Figure 1** shows how to replace regular AA or 9-volt batteries in the transmitter with 3-volt watch batteries.

If you connect two wires across the transmitter's activator button, you can have another sensor or switch activate the transmitter to alert you of an entry breech or that your valuables are being removed. Place a piece of tape over the transmitter button so that when the device is activated it will be on. **See Figure 2**.

ADAPTING THE CAR'S RECEIVER. You can also modify the car's radio receiver, which is on a circuit board in the car's body, for use as an alarm trigger. **See Figure 3.**

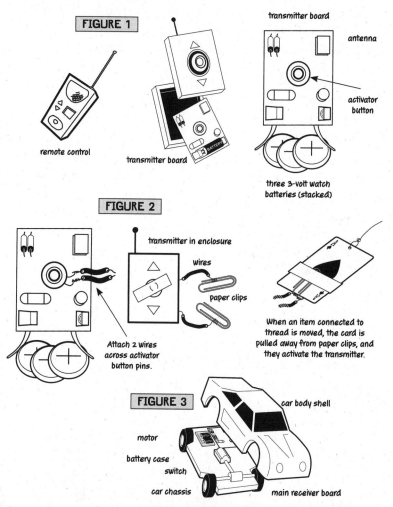

**FIGURE 1**

remote control

transmitter board

transmitter board

antenna

activator button

three 3-volt watch batteries (stacked)

**FIGURE 2**

transmitter in enclosure

wires

paper clips

Attach 2 wires across activator button pins.

When an item connected to thread is moved, the card is pulled away from paper clips, and they activate the transmitter.

**FIGURE 3**

car body shell

motor

battery case

switch

car chassis

main receiver board

Unlike the transmitter, the receiver must stay on to be able to operate, and this produces a small constant drain on the batteries. **Figure 4** illustrates how to modify the receiver for use with watch batteries using the same technique described for the radio transmitter.

If desired, the toy car motor can be used in an application of your own design. (*Sneaky Uses for Everyday Things* illustrated a Door Opener using a toy car.) The car motor is attached to the receiver with two connecting wires. If you physically remove the motor from the car body (either by unclipping or unscrewing it), you can use the receiver for more project applications. It's easy to connect the receiver's motor wires to other devices to activate them remotely.

**Figure 5** shows how the wires in the receiver that previously connected to the motor can be connected to other devices, like an LED or a buzzer for remote control.

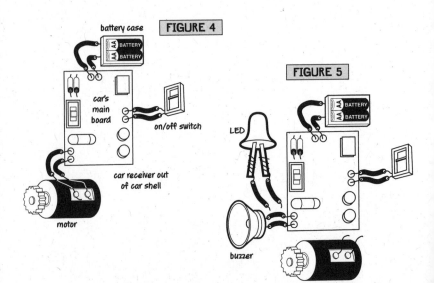

**FIGURE 4**

battery case

car's main board

on/off switch

car receiver out of car shell

motor

**FIGURE 5**

LED

buzzer

# SNEAKY FLYING DISK

Now it's time to make a sneaky flyer, similar to flying disc toys, using paper, scissors, and tape.

## What's Needed
▶ Scissors
▶ Paper, 8½ x 11 inches
▶ Transparent tape

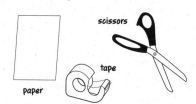

scissors

tape

paper

## What to Do
Cut eight 2-inch square pieces of paper as shown in **Figure 1**. Fold the top right corner of one square down to the lower left corner. See **Figure 2**. Then, fold the top left corner down to the bottom, as shown in **Figure 3**.

Repeat these two folds with the remaining seven squares. See **Figure 4**.

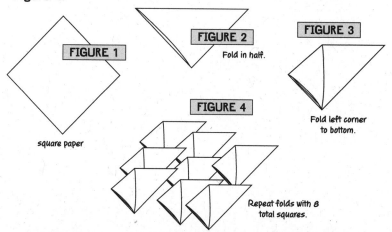

**FIGURE 1**

square paper

**FIGURE 2**

Fold in half.

**FIGURE 3**

Fold left corner to bottom.

**FIGURE 4**

Repeat folds with 8 total squares.

Insert one paper figure into the left pocket of another, as shown in **Figure 5**. Repeat, inserting the figures into one another until they form an eight-sided doughnut shape; see **Figure 6**. Apply tape as needed to keep the origami flyer together, and turn over, as shown in **Figure 7**.

Next, bend down the outer edge of the sneaky flyer to form a lip, as shown in **Figure 8**. This outer lip will cause the air to take a longer path over it, producing a Bernoulli effect.

Turn the device so the lip is bent downward. Throw the Sneaky Flying Disc with a quick snap of your wrist, and it should stay aloft for a great distance.

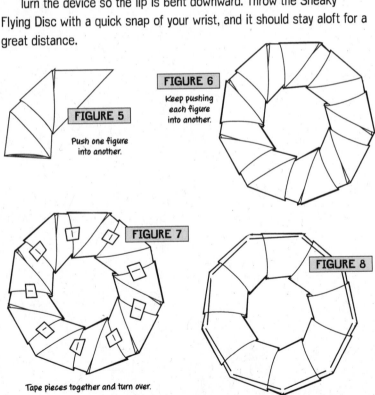

**FIGURE 5**

Push one figure into another.

**FIGURE 6**

Keep pushing each figure into another.

**FIGURE 7**

Tape pieces together and turn over.

**FIGURE 8**

Bend outer edge down for the Bernoulli effect.

# SNEAKY MINI BOOMERANG

You can use postcards, business cards, or cardboard food boxes to make a miniature, palm-size boomerang that actually flies and returns to you, for indoor fun.

## What's Needed

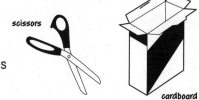

scissors

cardboard

- ▶ Scissors
- ▶ Cardboard from food boxes or postcards

## What to Do

Cut out the boomerang shapes shown in **Figure 1**. The boomerang wings can be any length between 2 to 4 inches. For optimal flight height and return performance, cut each wing of the boomerang 2½ inches long and ½ inch wide.

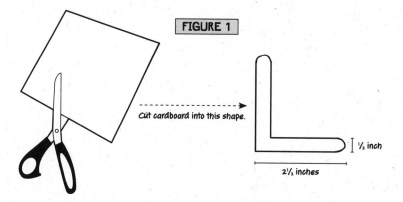

**FIGURE 1**

Cut cardboard into this shape.

½ inch

2½ inches

Set the Sneaky Mini-Boomerang on the palm of your raised hand with one wing hanging off. Tilt your hand slightly upward. With your other hand's thumb and middle finger ½ inch away, snap the outer boomerang wing. You'll discover (after a few attempts) that it will fly forward and return to you. See **Figure 2**.

**Note:** You must snap your finger with a strong snapping action to make the boomerang fly away and return properly, as shown in **Figure 3**.

Experiment with different hand positions and angles to control the boomerang's flight pattern.

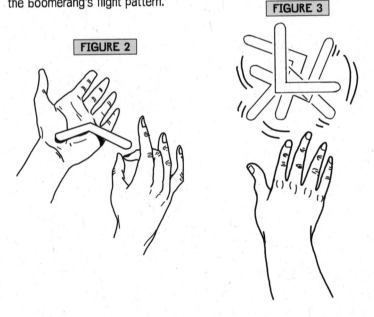

**FIGURE 2**

**FIGURE 3**

# SNEAKY ELECTRICAL GENERATOR

New energy sources are being found and refined every day, and you can demonstrate how industry, smaller businesses, and individuals take advantage of various forms of alternate energy sources. This project illustrates three methods that harness the power of wind, water, and steam to produce electricity. When a wire moves near a magnet, an electrical current is induced. Using this knowledge, you can create a Sneaky Electrical Generator with a toy motor.

## What's Needed

▶ Three large paper clips
▶ Electrical tape
▶ Toy car motor
▶ Pliers
▶ Voltmeter
▶ Wire (optional)

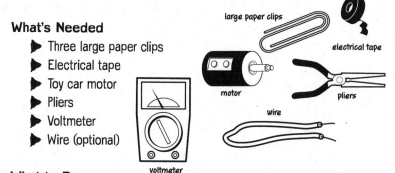

large paper clips

electrical tape

motor

pliers

wire

voltmeter

## What to Do

First, bend the three paper clips into the shapes shown in **Figure 1**. Paper clip one will act as a hand crank. The other two paper clips will act as propeller blades.

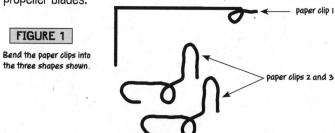

**FIGURE 1**

Bend the paper clips into the three shapes shown.

paper clip 1

paper clips 2 and 3

Next, wrap electrical tape around the shaft of the toy car motor. See **Figure 2**.

If the motor does not have wires on its two terminals, tape two 4-inch lengths of wire to them with tape.

Then, attach the first paper clip to the motor shaft and press it tight with pliers. Place the voltmeter on its lowest direct current (DC) setting and wrap the motor wires around its probes. Cranking the motor should cause the voltmeter to indicate a current has been generated, as shown in **Figure 3**.

Next, remove the first paper clip and press the other two paper clips onto the motor shaft as shown in **Figure 4**. Reshape the first paper clip to resemble the other two and press it onto the motor shaft also. See **Figure 5**.

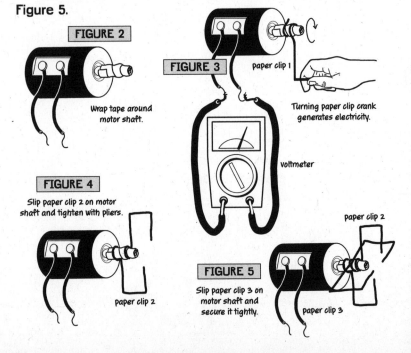

FIGURE 2
Wrap tape around motor shaft.

FIGURE 3
paper clip 1
Turning paper clip crank generates electricity.
voltmeter

FIGURE 4
Slip paper clip 2 on motor shaft and tighten with pliers.
paper clip 2

FIGURE 5
Slip paper clip 3 on motor shaft and secure it tightly.
paper clip 2
paper clip 3

Apply tape to all three paper clips to form propeller blades, as shown in **Figure 6**.

If you blow on the propeller or use a small hair dryer on it, the blades will turn and you will be harnessing wind power to generate electricity.

You can carefully hold the motor blades near a teakettle spout to harness steam power or place the propeller under a faucet's stream of running water to harness hydro power as shown in **Figure 7**.

**Note:** If you have a personal battery-powered fan, you can connect it to the voltmeter and spin its blades with your fingers to attain the same effect. See **Figure 8**.

**Figure 9** illustrates the internal parts of a wind turbine. Multiple wind turbines, called a wind farm, are shown in **Figure 10**.

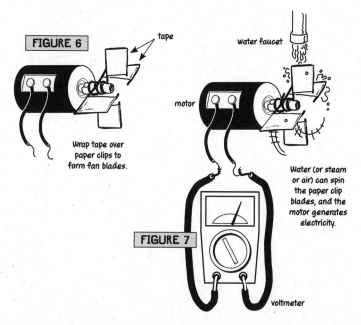

FIGURE 6

tape

Wrap tape over paper clips to form fan blades.

water faucet

motor

Water (or steam or air) can spin the paper clip blades, and the motor generates electricity.

FIGURE 7

voltmeter

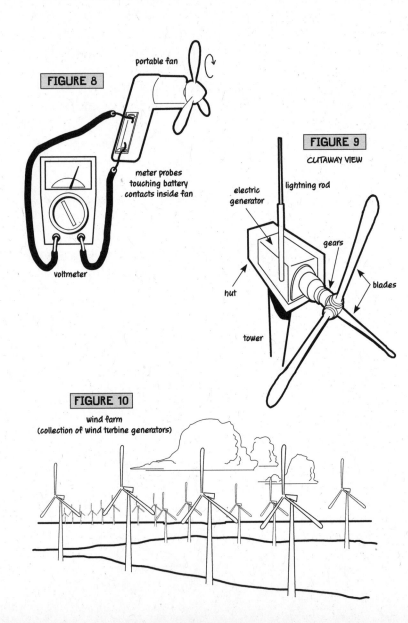

**FIGURE 8**

portable fan

meter probes
touching battery
contacts inside fan

voltmeter

**FIGURE 9**

CUTAWAY VIEW

lightning rod

electric
generator

gears

blades

hut

tower

**FIGURE 10**

wind farm
(collection of wind turbine generators)

# WILD, WILD VEST

The following project will illustrate how you can modify a favorite vest to access your favorite items and sneaky devices for a variety of purposes. You can mix and match different gadget sets according to your "mission." You're limited only by the availability of devices and your imagination.

The various vest-accessorizing examples that follow were selected for practicality. If desired, you can outfit your sneaky vest with a personal siren alarm, mini voice recorder, a retractable key ring, and a camera.

When you are traveling, the vest can be outfitted with a compass, mini poncho, pocket heater, mini survival kit, telescope, and a defensive repellent sprayer. You'll no doubt develop more personal creations that will express your mood and personality.

## What's Needed

- Vest
- Cloth that matches the vest's interior color
- Scissors
- Needle
- Thread
- Velcro strips or dots with sticky tape backing

## Optional:

- Personal alarm
- Retractable key ring
- Whistle

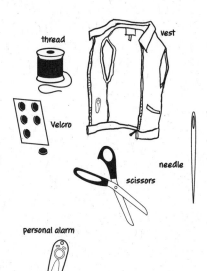

▶ Compass
▶ Mini mirror (from a compact makeup kit)
▶ LED blinking lights
▶ Camera
▶ Superthin, protective poncho
▶ Mini voice recorder
▶ Mini telescope

camera

compass

LED

## What to Do

Select a vest, preferably with a collar, made from sturdy material. A blue jean vest fits the bill perfectly. With some extra denim from a pair of old, discarded jeans, and a needle and thread, you can add sneaky pockets for personal items, such as a superthin, protective poncho.

For safety, apply Velcro strips to the inside of the vest and to a personal alarm. Select the type that includes a cord that hangs so when it's pulled, the alarm blasts an ultraloud squeal. Allow the cord to hang just below the inside waist for easy access, as shown in **Figure 1**.

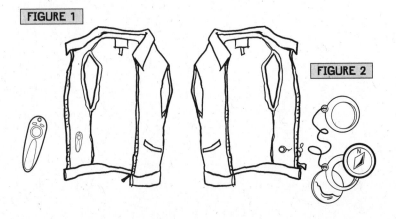

FIGURE 1

FIGURE 2

You can clip or sew a retractable key ring to the vest and attach a whistle, compass, mirror, and other devices to the vest, using the ring. Items may also be attached with Velcro dots applied directly to the vest. See **Figure 2**.

For fun, apply Velcro dots or strips to the vest so you can add LED blinking lights or your desired emblems and initials (ready-made from a store cut from spare fabric). This allows you to change your look when the mood strikes. See **Figure 3**.

The vest pockets can hold other valuable sneaky items, including a mini digital camera, superthin poncho, mini voice recorder, and mini telescope, as shown in **Figure 4**.

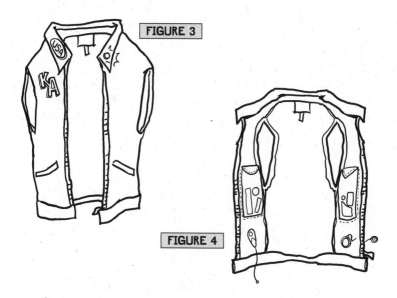

FIGURE 3

FIGURE 4

# SNEAKY TOY MODIFICATIONS

Who among us hasn't dreamed of having a power door opener as seen in sci-fi and spy movies? This project will show you how to use a small toy car to do the trick. A small wire-controlled car has enough power to push and pull a typical room door back and forth if you know the super-sneaky way to install it.

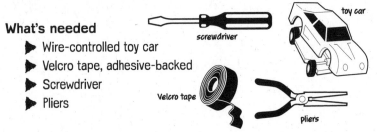

## What's needed
- ▶ Wire-controlled toy car
- ▶ Velcro tape, adhesive-backed
- ▶ Screwdriver
- ▶ Pliers

## What to do
This project requires a small wire-controlled toy car, not a radio-controlled version. This is to prevent the batteries from running down. (With a radio-controlled car, the remote control and the car's internal receiver have to be in the on mode, and this drains the batteries.)

First, remove the body shell from the toy car with a screwdriver. Then remove the front wheel and axle, as shown in **Figure 1**. Now, using the Velcro tape, attach the car near the bottom end of the door (see **Figure 2**).

Using the remote control, see if it can push the door open or closed. If not, reposition the car for more traction. When the proper position is found, you will be able either to move the door with your hand or let the car do it.

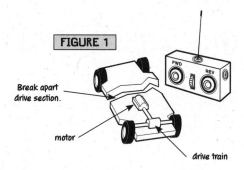

FIGURE 1

Break apart
drive section.

motor

drive train

FWD    REV

Optionally, you can break off the entire front part of the chassis so that it takes up less space and cover it with materials for a more appealing look. Mount the remote control outside the door as desired (see **Figure 3**).

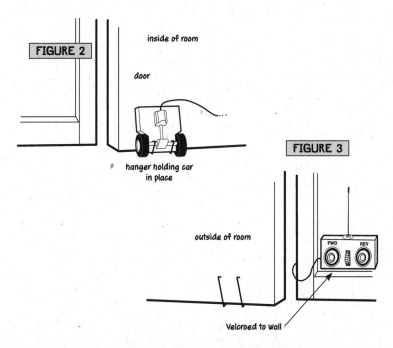

FIGURE 2

inside of room

door

hanger holding car
in place

FIGURE 3

outside of room

FWD    REV

Velcroed to wall

# SNEAKY FLOATING PHOTOS

With just a few simple items you can make a floating display box for your favorite small photographs. It will keep admirers in awe of how it defies gravity.

## What's Needed

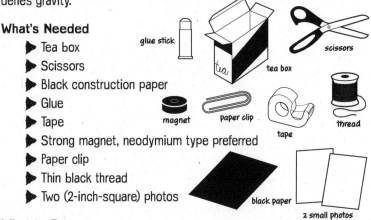

glue stick

scissors

tea box

magnet    paper clip    tape    thread

black paper

2 small photos

- ▶ Tea box
- ▶ Scissors
- ▶ Black construction paper
- ▶ Glue
- ▶ Tape
- ▶ Strong magnet, neodymium type preferred
- ▶ Paper clip
- ▶ Thin black thread
- ▶ Two (2-inch-square) photos

## What to Do

Unfold the tea box and cut off the top and side flaps. See **Figure 1**. Lay it flat and cut a piece of black construction paper to cover the inside of the tea box. Glue the paper to the cardboard and let it dry. Then cut slits in the paper to match the ones on the tea box. See **Figures 2** and **3**.

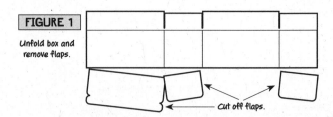

**FIGURE 1**

Unfold box and remove flaps.

◀── Cut off flaps.

Fold the tea box back into shape and glue it securely. The dark interior of the box will be a showcase for your photos. Tape the magnet to the inside top of the box and cover it with a small piece of black paper, as shown in **Figure 4**.

Tie a 3-inch length of thread to the end of the paper clip. Then, cover the paper clip with the two photos and tape them together, as shown in **Figure 5**.

**FIGURE 2**

Cut black paper to fit box dimensions and glue together.

**FIGURE 3**

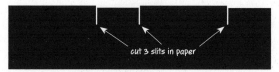

cut 3 slits in paper

Cut 3 slits in paper to match ones on tea box.

**FIGURE 4**

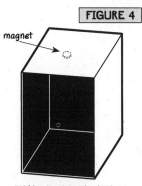

magnet

Fold box together, glue in place,
and tape magnet under top section.

**FIGURE 5**

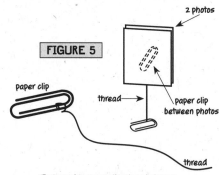

2 photos

paper clip

thread

paper clip between photos

thread

Tie thread to paper clip, place between
photos, and tape photos together.

Tape the loose end of the thread to the bottom center inside the box so the photos, when lifted to the top, have a ½-inch gap between them and the top of the box. Lift the photos to the top and let go. Since the paper clip inside them is attracted to the magnet above, they will float in midair, as shown in **Figure 6**. If necessary, adjust the length of the thread for a proper fit. When placed on a shelf, the black thread will be virtually invisible against the black backdrop and the photos will appear to defy gravity.

**FIGURE 6**

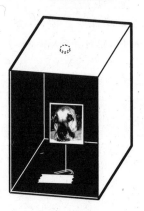

Tape thread to bottom of box,
and the photo will float like magic.

# Recommended Links

See www.sneakyuses.com for *Sneaky Uses* articles, how-to videos, and updates.

## Science and Technology Sites
Community of Science: www.cos.com
Education Freebies: www.thehomeschoolmom.com
*Science* Magazine: www.sciencemag.org
Science Master: www.sciencemaster.com
Siemens Foundation: www.siemens-foundation.org
U.S. Government Science Grants: www.science.doe.gov/grants
AMA Science: www.sciencetoymaker.org
By Kids For Kids: www.bkfk.com
Funology: www.funology.com
About.com: www.about.com
*Popular Science:* www.popsci.com
*Popular Mechanics:* www.popularmechanics.com
Rube Goldberg: www.rube-goldberg.com
The F.U.N. Place: www.thefunplace.com

## Science Project Sites
Sneaky Uses for Everyday Things: www.sneakyuses.com
Instructables: www.instructables.com
MakerBot Industries: www.makerbot.com
Exploratorium: www.exploratorium.edu
Arvind Gupta Toys: www.arvindguptatoys.com
Make Projects: www.makeprojects.com

Science Hobbyist: www.amasci.com
Build-It-Yourself: www.build-it-yourself.com
DoItYourself: www.doityourself.com
Howtoons: www.howtoons.net
Discover Circuits: www.discovercircuits.com
Kids Invent: www.kidsinvent.com
Make-Stuff: www.make-stuff.com
Ready-Made Plastic Trays: www.ready-made.com
Wacky Uses for brand Name Products: www.wackyuses.com
HowStuffWorks: www.howstuffworks.com
Science Project: scienceproject.com
Science Toys: scitoys.com

### Gadgets and Parts Sites
Think Geek: www.thinkgeek.com
Nuts and Volts: www.nutsvolts.com
All Electronics Corporation: allelectronics.com
Ramsey: www.ramseyelectronics.com
Digi-Key Corporation: www.digikey.com
Jameco Electronics: www.jameco.com
Miniatronics: www.miniatronics.com
Fun Gizmos: store.fungizmos.com
johnson-smith.com
Jameco Robot Store: www.jameco.com
Smart Planet: www.smartplanethome.com
Spy-Gear: spy-gear.net
Edmund Scientific: www.scientificsonline.com
Super Bright LEDs: www.superbrightleds.com

## Microcontroller Kit and Tutorial Links

Make Arduino: makezine.com/arduino/

Arduino Tutorial: www.ladyada.net/learn/arduino/

Adafruit Industries: adafruit.com

Sparkfun Electronics: www.sparkfun.com

## Other Web Sites of Interest

*Craftmag:* www.craftmag.com

*Craft:* craftzine.com

Fashioning Technology: www.fashioningtech.com

*Wired:* www.wired.com

Freecycle: www.freecycle.org

Geek Technique: www.geektechnique.org

Choose2Reuse: choose2reuse.org

*Nails* Magazine: www.nailsmag.com

*Nailpro:* Nailpro.com